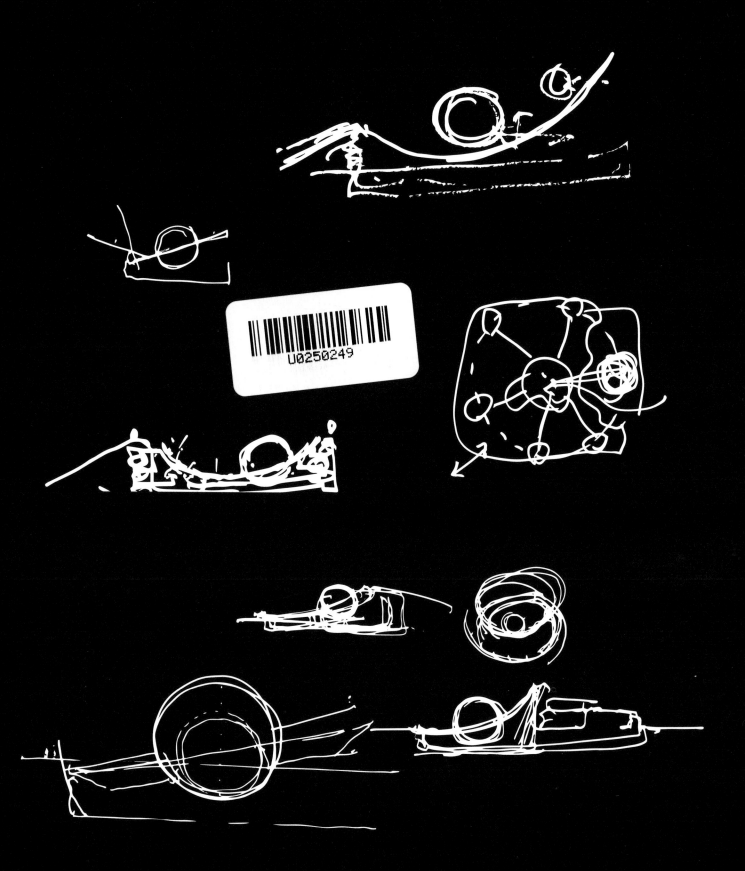

ARCHITECTURE OF THE COSMOS
Shanghai Astronomy Museum

宇宙的构筑
上海天文馆

Ennead Architects LLP
New York | Shanghai

艺艾德建筑设计事务所
纽约 | 上海

同济大学出版社·上海
TONGJI UNIVERSITY PRESS · SHANGHAI

图书在版编目（CIP）数据

宇宙的构筑：上海天文馆：英汉对照 / 美国艺艾德建筑设计事务所著. -- 上海：同济大学出版社，2023.8

ISBN 978-7-5765-0532-0

Ⅰ.①宇… Ⅱ.①美… Ⅲ.①天文馆 – 建筑设计 – 研究 – 上海 – 英、汉 Ⅳ.①TU244.6

中国版本图书馆CIP数据核字(2022)第242501号

群岛ARCHIPELAGO是专注于城市、建筑、设计领域的出版传媒平台，由群岛ARCHIPELAGO策划、制作、出版的图书曾荣获德国DAM年度最佳建筑图书奖、政府出版奖、中国最美的书等众多奖项；曾受邀参加中日韩"书筑"展、纽约建筑书展（群岛ARCHIPELAGO策划、出版的三种图书入选为"过去35年中全球最重要的建筑专业出版物"）等国际展览。
群岛ARCHIPELAGO包含出版、新媒体与群岛BOOKS书店。
archipelago.net.cn

宇宙的构筑
上海天文馆

艺艾德建筑设计事务所 著

出版人：金英伟
策　划：秦蕾 / 群岛ARCHIPELAGO
特约编辑：辛梦瑶
责任编辑：王胤瑜　晁艳
责任校对：徐逢乔
版式设计：Thomas Wong　Aislinn Weidele　付超
版　次：2023年8月第1版
印　次：2023年8月第1次印刷
印　刷：上海雅昌艺术印刷有限公司
开　本：787mm × 1092mm　1/12
印　张：15+1/3
字　数：386 000
书　号：ISBN 978-7-5765-0532-0
定　价：198.00元
出版发行：同济大学出版社
地　址：上海市四平路1239号
邮政编码：200092
网址：http://tongjipress.com.cn

本书若有印装质量问题，请向本社发行部调换。
版权所有 侵权必究

ARCHITECTURE OF THE COSMOS
Shanghai Astronomy Museum

by Ennead Architects LLP

ISBN 978-7-5765-0532-0

Publisher: Jin Yingwei

Initiated by: Qin Lei/ARCHIPELAGO

Contributing Editor: Xin Mengyao

Editors: Wang Yinyu, Chao Yan

Proofreader: Xu Fengqiao

Graphic Designers: Thomas Wong, Aislinn Weidele, Fu Chao

Published in August 2023, by Tongji University Press,
1239, Siping Road, Shanghai 200092, China
http://tongjipress.com.cn

archipelago.net.cn

All rights reserved
No part of this book may be reproduced in any
manner whatsoever without written permission from
the publisher, except in the context of reviews.

Contents   目录

008   **Foreword**
前言
Ye Shuhua
叶叔华

012   **Introduction: Welcome from the Shanghai Astronomy Museum**
序章：上海天文馆欢迎您
Shanghai Science and Technology Museum
上海科技馆

024   **Connecting Humans to the Universe**
连接人类与宇宙
Thomas Wong
托马斯·黄

042   **Site**
场地

056   **Concept**
概念

078   **Realization**
实现

154   **Engineering and Construction**
工程与建造

182   **Acknowledgements**
致谢

A view of the Earth from the Apollo 11 spacecraft in the Mare Smythii on the moon, July 1969

1969年7月，阿波罗11号宇宙飞船拍摄到的地球从月球"史密斯海"升起的景象

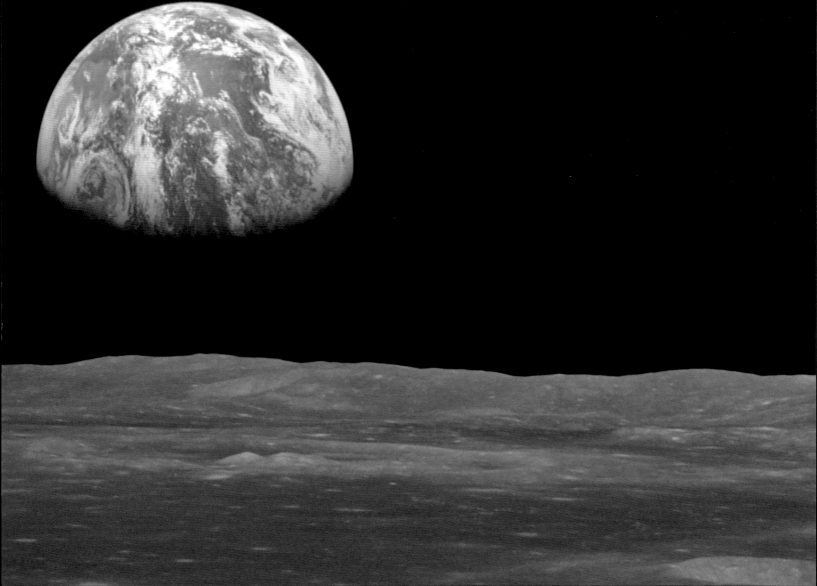

Foreword

A beautiful planetarium has appeared on the banks of the Dishui Lake. Its shape outlines abstract celestial orbits, and its interior exhibition is even more exciting.

The Beijing Planetarium was built much earlier, in 1957. When can people in Shanghai have their own planetarium? This dream actually started 45 years ago. 30 years ago, I myself began to call for the construction of the Shanghai Planetarium. Unfortunately, due to the limited economic conditions of the country at that time, the contents of a planetarium could only be partially displayed at the Shanghai Science and Technology Museum. 10 years ago, taking advantage of the momentum of the World Expo, I once again appealed to the municipal government to build a planetarium in Shanghai. This time, the government responded very quickly. Secretary Yu Zhengsheng and Mayor Han Zheng personally authorized an effort to conduct research and implement a construction plan as soon as possible.

I want to thank the Shanghai Science and Technology Museum for taking on this glorious task at this critical moment. Chen Mingbo, deputy director of the Shanghai Municipal Science and Technology Commission and Secretary of the Party Committee of the Shanghai Science and Technology Museum, immediately organized a capable preparatory team and coordinated with relevant comrades of the Shanghai Astronomical Observatory of the Chinese Academy of Sciences. The solicitation of international designers, and the conceptual plan for the planetarium to be located in Lingang New City were subsequently determined. The project was formally approved in January 2014, and construction started in November 2016, almost completed by now. In my lifetime, I have personally experienced the whole construction process of the world's largest planetarium, watched it grow and lift off the ground with my own eyes. I am very pleased.

The planetarium is a unique popular science venue. The most wonderful thing about it is that it can reproduce a realistic starry sky and create an atmosphere of space exploration. In this carefully designed and mysterious space, the Shanghai Astronomy Museum has applied a variety of high-tech displays, designed different levels of exhibit experiences for an array of tourists, and displayed the latest developments in modern astronomy in an easy-to-understand manner.

The planetarium is an important place for shaping a real understanding of the universe and helping to improve the scientific literacy of the general public. I sincerely hope that the Shanghai Astronomy Museum can truly become a popular science venue that is "always open and always new." The development of modern science is changing with each passing day, and the achievements in aerospace and astronomy research are endless. Our museum must not rest on the knowledge of the past. Many of the current digital displays allow easy updates of the content. Therefore, the education team needs to keep track of the latest astronomical developments and display the latest knowledge of astronomy in our museum.

I am very pleased that the Shanghai Astronomy Museum also established an astronomical research center. This is also an important measure to maintain its academic stature and keep it current and new. I sincerely hope that this center can be combined with the features of the museum itself to support first-class scientific research while maintaining the highest level of exhibition and scientific display.

I sincerely hope that the Shanghai Astronomy Museum can truly become a leader in China's astronomy science education by attracting and training more astronomy enthusiasts and workers, and contributing to the future development of China's astronomy industry.

Ye Shuhua
Shanghai
March 2021

前言

美丽的滴水湖畔，出现了一座美丽的天文馆。它优美的外形勾勒出抽象的天体轨道，它内部的展示体验更是精彩纷呈。

北京天文馆早在1957年就已诞生，上海人什么时候也能拥有自己的天文馆？这个梦想其实早在45年前就已开始了。30年前，我自己也开始呼吁建设上海天文馆，遗憾的是，那个时候国家的经济条件有限，天文馆的内容还只能在上海科技馆中作部分展现。10年前，借着世博会的东风，我再次向市政府呼吁建设上海天文馆。这一次，政府的反应非常迅速，当时的俞正声书记、韩正市长亲自批示，要求快速展开调研，尽快落实建设计划。

我要感谢上海科技馆，在这个关键时刻接下了这个光荣的任务。当时的上海市科委副主任兼上海科技馆党委书记陈鸣波同志，立即组织了精干的筹备班子，协同中科院上海天文台的有关同志，经过了扎实细致的专家咨询、选址论证、内容策划、建筑方案和展陈设计国际竞赛等多个步骤，最终确定了落户临港新城的上海天文馆概念方案。2014年1月正式立项，2016年11月开工建设，到今天，上海天文馆已经基本建成，很快就将迎来开馆试运行。能在有生之年，亲历这一世界最大天文馆的建设全过程，亲眼看着它孕育成长，呱呱坠地，我甚为欣慰。

天文馆是一种十分独特的科普场馆，它最奇妙之处就是能够再现一个逼真的星空，还能创造一个太空漫游的氛围。在这种精心设计的神秘空间里，上海天文馆应用了多种高科技的展示手段，针对不同的游客设计了不同层次的体验项目，尽可能地用通俗易懂的方式来展现现代天文学发展的最新成果。

天文馆是一个塑造正确宇宙观的重要场所，有助于提升广大市民的科学素质。我也一直衷心地希望上海天文馆能够真正成为一座"常开常新"的科普场馆。现代科学的发展日新月异，航天和天文学研究的成果层出不穷，我们的天文馆绝对不能躺在以往的知识上睡大觉。现在的很多展示手段都是多媒体，可以十分方便地更新内容，因此需要我们天文馆的教育团队时刻注意跟踪最新天文发展，随时将最新的天文新知展现在我们的天文馆里。

我很高兴天文馆不久前也成立了天文研究中心，这也是上海天文馆能够保持学术水平，保持常开常新的重要举措，衷心希望这个研究中心能够结合天文馆自身的特点，做出富有特色的一流科研成果，同时指导整个天文馆的展示水平始终保持最高的水准。衷心祝愿上海天文馆未来能真正成为中国天文科普教育的领跑者，能够吸引和培养更多的天文爱好者和天文工作者，为中国天文事业的未来发展贡献自己的力量。

叶叔华
上海
2021年3月

Ye Shuhua, born in 1927 in Guangzhou, is an astronomer known for her precise measurement of Universal Time in the 1960s. She was elected to the Chinese Academy of Sciences in 1980, served as director of the Shanghai Astronomical Observatory from 1981 to 1993, and was vice president of the International Astronomical Union from 1988 to 1994. She has been the honorary president of the Chinese Astronomical Society since 1989. The asteroid 3241 Yeshuhua is named for her.

叶叔华，中国天文学家，1927年生于广州，20世纪60年代即以精确测量世界时而闻名。叶先生于1980年当选为中国科学院院士，1981—1993年任中国科学院上海天文台台长，1988—1994年担任国际天文学联合会副主席，并自1989年起担任中国天文学会名誉理事长。3241号小行星 Yeshuhua 就是以叶先生的名字命名的。

We have a hunger of the mind which asks for knowledge of all around us, and the more we gain, the more is our desire; the more we see, the more we are capable of seeing.

我们内心有一种渴望，渴望了解周围的一切，我们得到的越多，欲望就越多；我们看到的越多，就越能发现更多。

— Maria Mitchell, American Astronomer and Naturalist, 1818—1889
玛丽亚·米切尔（1818—1889），美国天文学家、博物学家

Introduction: Welcome from the Shanghai Astronomy Museum
序章：上海天文馆欢迎您

By Shanghai Science and Technology Museum
上海科技馆 撰文

The philosopher Immanuel Kant once said: "Two things fill the mind with ever new and increasing admiration and awe, the more often and steadily we reflect upon them: the starry heavens above me and the moral law within me." The unity of man and nature has been a crucial concept in Chinese philosophy since ancient times. With their persistent curiosity, the Chinese people have unceasingly observed the starry skies and endlessly explored the mysteries of the universe: Taosi, the remains of China's earliest observatory in Xiangfen County, Shanxi, dates to 2000 BC; Dengfeng Observatory was developed by the astronomer Guo Shoujing for geodetic astronomy in the 13th century, the Yuan dynasty; the Chinese Astronomical Society was established a hundred years ago; and, in recent years, the Dark Matter Particle Explorer (DAMPE), or Wukong, was launched to explore dark matter, and Chang'e 5 took off to retrieve samples from the moon.

Today, the largest planetarium in the world that seeks to arouse curiosity about the universe and fulfill astronomical pursuits has opened in Lin-gang Special Area of the China (Shanghai) Pilot Free-Trade Zone on the coast of the East China Sea, the frontier of China's reform and opening-up policy.

Where the planetarium is located is a magical place with the captivating name Dishui Lake, or Dripping Lake. Legend has it that a drop of water fell from the sky and expanded and evolved into a lake. Its ripples spread out and became roads that encircle the lake. Thus, Lin-gang Special Area (which translates to "the area that sits beside a port") was born. The name refers to Yangshan Port, a busy international seaport in Shanghai. This free-trade zone beside the port is an embodiment of the sheer enthusiasm and ambitions of the people of Shanghai to build the city into an international metropolis.

Among other new developments in the city that celebrate scientific and technological advancements, the Shanghai Astronomy Museum stands out as an architectural icon with a unique program. The re-creation of the fascinating night sky and vivid representation of the mysterious universe allows visitors' minds to roam to other worlds, indulging their curiosity about the vastness of the universe. This immersive experiential endeavor encourages visitors to explore the relationship between man and nature.

The Shanghai Astronomy Museum is integral to the Shanghai Science and Technology Museum, which forms a

哲学家康德说过："世界上有两种东西能震撼人们的心灵，一是我们心中崇高的道德标准，二是我们头顶灿烂的星空。"天人合一，也是中国文化自古以来很重要的一个哲学概念。从中国考古学家在山西襄汾陶寺古城遗址发现公元前2000多年中国最早的天文台遗址，到公元13世纪元代天文学家郭守敬建立登封古观象台进行大地天文测量；从100年前中国天文学会正式诞生，到中国今天的"悟空"卫星升空探索暗物质、"嫦娥五号"飞船从月球取回土壤，从古至今，中国人从未失去对浩瀚星空的好奇和观察，从未停止对宇宙奥秘的思考和探索。

如今，一座引领人们探究宇宙真相、追寻天文梦想的世界最大天文馆已经在东海之滨——中国改革开放的前沿阵地，上海自由贸易试验区临港新片区建成开放。这是一个神奇的地方，有着梦一样的名字——滴水湖。传说，天上落下了一滴水，水滴成了湖，湖水涟漪向外扩散，变成了一圈一圈的环湖道路，造就了一座临港新城，城市名中的那个港，叫洋山港，上海人把它建设成举世瞩目的国际海运大港。国际港、自贸区、临港城，这里承载了上海人建设国际大都市的豪情壮志。

就在这片建设科技创新之城的热土上，上海天文馆成了一个新地标。上海天文馆是一座特别的科普场馆，它将再现深邃的星空，展示神秘的宇宙世界，带领观众暂时脱离尘世，激起每一个人对星空和宇宙的敬重，思考天与人的关系。

上海天文馆也是世界博物馆行业排名靠前的上海科技馆

2002 2015 2020

Dishui Lake 滴水湖

unified cluster of three museums that are among the world's elite. The Shanghai Science and Technology Museum is one of the largest comprehensive exhibition spaces dedicated to science and technology in China. Focusing on the themes of "Nature," "Human," and "Technology," it serves as a portal for education, research, and knowledge exchange. Ranked as a top-tier national museum and a 5A national tourist attraction in China, the museum champions interdisciplinary methodologies to explain natural science concepts in engaging ways that entice visitors to explore.

In April 2015, the first branch of the Shanghai Science and Technology Museum, the 45,000-square-meter Shanghai Natural History Museum, was completed and opened in Shanghai Jing'an Sculpture Park. The Shanghai Astronomy Museum operates as its sister entity; its completion marks the final phase of the Shanghai Science and Technology Museum cluster, intended for popular science education.

Conception

A planetarium in modern times is a place where sophisticated optical instruments are used to assimilate educational experiences that teach astronomy and all related sciences. The world's first projection planetarium opened in 1925 at the Deutsches Museum in Munich. Since then, projection planetariums have become immensely popular across the world. Not long after the founding of the People's Republic of China, the country's first planetarium, the Beijing Planetarium, opened in September 1957. It continued to expand its exhibition program and was moved to new premises in 2004.

This acclaimed planetarium became the biggest hub for astronomy in China, inspiring generations to pursue astronomical studies. For a long time, however, the Beijing Planetarium remained the only large planetarium in China and could not meet the increasing demand for astronomical knowledge nationwide. Hence, Shanghai, an equally important city, was pressured to build its own.

As early as the 1970s, leaders such as Li Xiannian and Gu Mu initiated a proposal to build a planetarium in Shanghai. In the 1990s, four academicians — Hsi-teh Hsieh, Tan Jiazhen, Weng Shilie, and Ye Shuhua — submitted a proposal, but for reasons that remain unclear, it never saw the light of day.

During the Shanghai World Expo in July 2010, the Academician of Chinese Academy of Science and an esteemed scientist from the Shanghai Astronomical Observatory, Ye Shuhua, sent a formal letter to the city government, once again asking leaders to seriously consider the notion of raising the awareness of astronomy by building a sister planetarium in Shanghai. In her proposal, she wrote: "Deep space exploration will become the subject of competition for technological supremacy in various countries in the 21st century. Hence setting aside our priorities to build a Shanghai planetarium would be a timely endeavor to promote astronomical knowledge among the public by supplementing an informal education program in astronomy for young people. The venue must be exciting, stimulating, and grand so that an introduction to such a specialized field would be entertaining, rather than institutional."

Her proposal won the attention and support of the Shanghai government. Yu Zhengsheng, then the Secretary of Municipal Party Committee of Shanghai, and Han Zheng, then the mayor, instructed relevant departments to conduct surveys and organize research groups to study how this form of scientific education might successfully reach the masses. In 2012, the municipal government officially commissioned the Shanghai Science and Technology Museum to implement the construction of the Shanghai Astronomy Museum, with the Shanghai Astronomical Observatory under the Chinese Academy of Sciences providing professional guidance. After numerous considerations for preferred locations, it was decided that the Shanghai Astronomy Museum would be built within the city park area adjacent to the North Huanhu 3rd Road on the northern side of Dishui Lake Station on Metro Line 16 in the Lin-gang Special Area in Pudong New Area.

Shanghai Science and Technology Museum, established 2001, is one of the largest comprehensive science and technology museums in China
上海科技馆成立于2001年，是中国规模最大的综合性科学技术类博物馆之一

"三馆合一"集群建设的重要组成部分。上海科技馆是中国规模最大的综合性科学技术类博物馆之一，以"自然·人·科技"为主题，融展示与教育、收藏与研究、合作与交流、休闲与旅游于一体，以学科综合的手段及寓教于乐的方式诠释自然与科学技术知识，引发观众探索自然与科技奥秘的兴趣，也是国家一级博物馆和5A级旅游景区。

2015年4月，位于上海静安雕塑公园的上海自然博物馆建成开放，它是上海科技馆的第一个分馆，建筑面积4.5万平方米。上海天文馆建成之后将成为第二个分馆。从此，上海科技馆"三馆合一"的科普教育集群正式形成。

缘起

现代意义上的天文馆，是一种使用光学天象仪来模拟和演示真实星空的场所，世界上第一座天文馆1925年诞生于德意志博物馆，广受欢迎，从此在世界各地普及开来。中华人民共和国建立之后，很快也建立了中国第一座天文馆，1957年9月，北京天文馆正式对外开放，2004年又完成了新馆建设。

北京天文馆建成之后赢得了广泛的社会声誉，成为国内最主要的天文科普基地，激发了几代人对天文科学的热爱。然而长期以来，中国的大型天文馆却只限于北京天文馆，远远无法满足国人对于天文科普的需求。作为同样的国际大都市，上海也在梦想着建立自己的大型天文馆。

早在20世纪70年代，李先念、谷牧等中国领导人就曾经批示建设上海天文馆。20世纪90年代，谢希德、谈家桢、翁史烈和叶叔华四位院士再次提议建设上海天文馆。然而，由于各种原因，上海天文馆之梦始终未能付诸实现。

2010年7月，在举世瞩目的世博会举办期间，上海天文台的中国科学院院士叶叔华再次向上海市政府郑重致函，建言建设上海天文馆。她在建议中提到："21世纪深空探测将成为各国科技竞赛场……兴建上海天文馆，对广大市民进行天文科学知识普及推广，补充青少年天文科学方面的非正规教育，并使其成为上海又一科技人文景观，休闲之余有所学得，更相得益彰。"

此建议迅速得到了上海市的关注和支持，时任市委书记俞正声、市长韩正批示请相关部门组织调研。2012年，上海市政府正式委托上海科技馆承担上海天文馆的建设任务，中科院上海天文台负责提供专业支持。经过慎重的选址比较，上海天文馆最终确定落户在上海浦东新区的临港新城，位于地铁16号线滴水湖站北侧紧挨环湖北三路的城市公园区域内。

Fruition

In September 2014, international bids were solicited for the design of the Shanghai Astronomy Museum, and the U.S. firm Ennead Architects was awarded the contract. Ennead used astronomical concepts such as gravitational pull and orbital motion for the design themes of the main building of the Shanghai Astronomy Museum and its surrounding structures. The building's three principal forms — the Oculus, the Inverted Dome, and the Sphere — act as "celestial bodies," a fundamental concept in celestial mechanics. Through scale, form, and the manipulation of light, the building resembles a mysterious time machine. On a bigger scale, the building's orbital design skillfully echoes the ring roads that encircle Dishui Lake, creating a visual dynamic that appears to animate the structure.

In September 2015, the feasibility study report for the Shanghai Astronomy Museum project was officially approved. On November 8, 2016, ground was broken. The construction team went through countless challenges but resolved them to make their shared dream come true. The long days and nights left a mark on every member of the construction team, and their experiences became valuable assets.

The Shanghai Astronomy Museum was executed with the highest of standards, with innovative solutions and the strictest requirements throughout the process. The Shanghai municipal government set the highest mark of excellence for "building the world's top-class planetarium" from the start. Wang Lianhua, Secretary of the Shanghai Science and Technology Museum Party Committee, insisted that the construction of the planetarium be in line with the most advanced facilities in the world. Specifically, he required that tradition and history of Chinese astronomy be incorporated, and modern technologies applied, so that both global astronomical history and regional scientific development be reflected, and both international and Chinese perspectives of astronomical pursuits be contemplated. His mission statement for the museum was that it be a scientific educational site for exhibition and research. He wanted it to be a venue for insights and aspirations while functioning as an urban public and cultural space.

The complex technical processes and unconventional structural design of the Shanghai Astronomy Museum posed challenging construction methodologies. To build a world-class astronomy museum, the construction team knew that it had to take on research and development in all aspects of construction; for example, proactively adopting new materials and assembly methods and advancing their integration of key technologies in energy-saving solutions, management methods, and green landscape construction.

The team also took a lead role in applying Building Information Modeling (BIM) technology to the life-cycle management of the museum during its construction. Throughout the process of preliminary design, detailed construction drawings, on-site construction preparation, project implementation, and operations, the application of BIM assisted with the smooth coordination between all consultants and trades. With the full engagement of BIM, errors in engineering design and construction were minimized, and the project was completed on time, with the highest quality. Due to the outstanding performance, the Shanghai Astronomy Museum project was selected as a pilot project for BIM technology application in Shanghai, winning the Excellent BIM Application Award for Culture and Sports in the 2015 Innovation Cup BIM Application Design Competition, as well as the honorary title of "Featured Project" in the 2019 Shanghai Key Project Merit Competition.

In September 2019, the shell and core of the Shanghai Astronomy Museum were completed; then began the full-scale interior architecture and furnishing procurement, along with the research and development to produce the museum's immersive exhibits. In April 2021, the interior architecture was completed, and all commissioned exhibits had been installed. In July 2021, this monumental effort, the Shanghai Astronomy Museum, was officially opened to the public, designated as another landmark, together with the nearby starry sky theme park, in Lin-gang Special Area.

建设

2014年9月，上海天文馆建筑方案进行国际招标，美国艾艾德建筑设计事务所的设计方案中标。该方案将"引力和轨道"等天文概念充分应用于上海天文馆主建筑及其周边场地的设计主题。三个独特的建筑结构体组成了作为天体力学基本概念的"三体"，同时巧妙设计光影效果，使得整个建筑体成为一座巨大而神秘的时间机器。在大空间范围内，天文馆建筑体的轨道设计还巧妙地与滴水湖畔的环形道路形成呼应，强烈的动感能量激发了整个建筑的活力。

2015年9月，上海天文馆建筑工程项目可行性研究报告正式获批。2016年11月8日，上海天文馆正式开工建设。几年来，天文馆从无到有，从小到大，全体建设人员克服了一个个困难，解决了一个个难题，一步步地把梦想变成了现实。建设过程中的点点滴滴，都在每位参建人员的心里留下深刻的印记，成为生命中一笔宝贵的财富。

上海天文馆从立项之初就始终坚持高起点、高要求。上海市政府一开始就提出了"建设国际顶级天文馆"的高标准，上海科技馆党委书记王莲华同志始终要求对标国际最先进的天文馆建设理念，要求天文馆建设做到继承传统与应用现代科技相结合、演绎天文历史全貌与彰显区位人文地理相结合、国际化的科学视野与中国人的天文探索相结合，既要考虑到天文馆作为科普教育基地在展示、教育、科研等方面的需要，也应该考虑城市公共文化场所功能定位的新要求、新思路。

上海天文馆建筑设计独特，技术工艺复杂，异形结构多，施工难度高。为建设世界一流的天文馆，项目建设团队以科研先行，大胆探索，积极采用新材料、新工艺，对设计和施工中涉及节能方案、管理模式、绿地景观建设等多方面的关键技术开展了创新课题研究，为项目的顺利实施提供了技术支持。

上海天文馆在建设过程中，还率先将BIM技术应用于建筑体全生命周期管理。在项目初步设计、施工图设计，以及施工准备、实施、运营的全过程中，通过BIM技术应用更好地协同各参建方，发挥BIM技术优势，并使工程设计和施工的错误降到最少，确保按时优质地完成项目建设。上海天文馆项目因为在BIM技术上的突出表现，获选为上海市BIM技术应用试点项目，荣获了2015年"创新杯"建筑信息模型（BIM）设计大赛"文化体育类优秀BIM应用奖"，以及2019年上海市重点工程实事立功竞赛"特色项目"等荣誉称号。

2019年9月，上海天文馆建筑正式竣工，随即全面展开了展示工程的装饰布展施工，同时展开了众多展品展项的招标采购和研发生产。2021年4月，装饰布展工程基本完成，展品展项基本完成安装调试。2021年7月，一座精彩的大型天文馆正式面向公众开放，它与周边的星空之境主题公园一道，成为临港新片区又一张亮丽的名片。

愿景和使命

上海天文馆得天独厚，它拥有世界规模最大的天文主题建筑体，也拥有一个能最大程度全面反映人类探索宇宙的成果和历程的大型展厅。为此，上海天文馆响亮地提出了"塑造完整宇宙观，激发公众的好奇心，鼓励公众感受星空、理解宇宙、思考未来"的愿景和使命。

他们希望通过精彩的场馆展示体验和丰富的教育活动，来帮助到访的观众清晰、完整地建立起对我们所在这个宇宙的总体认识，了解宇宙运行的基本规律，并通过讲述人类探索这些奥秘的故事，帮助人们真正理解什么是科学精神，什么是科学方法和科学的思维方式。

好奇心是人类探索宇宙的原动力，相对于观众在天文馆中学到的知识，上海天文馆更看重的是通过参观体验，激发出观众内心深处潜藏着的对宇宙、对自然的好奇心，这种好奇心才是推动科学发展的真正动力，并吸引更多的人孜孜不倦地去探寻科学真理。

星空是连接人和宇宙的桥梁。没有星空，人类将无从知晓宇宙的存在；没有星空，我们将无法窥探遥远天体运行的奥秘。如今，对上海的公众而言，星空已经成了一个传说，城市的发展使我们已经难以看到真正的星空。上海天文馆将为公众

Vision and Mission

The Shanghai Astronomy Museum is unique in that it has the world's largest astronomy-themed building that houses an exhibition hall that fully reflects the achievements and history of humanity's exploration of the universe. As such, the Shanghai Astronomy Museum sees its vision and mission as "assisting the establishment toward a comprehensive understanding of the universe, arousing the curiosity of the public, and encouraging them to perceive the galaxy of stars, understand the phenomenality of the universe, and contemplate the future."

By hosting extraordinary exhibitions and an abundance of academic activities, as well as presenting the stories of how humanity explores the mysteries of the universe, the Shanghai Astronomy Museum aims to help visitors have a clear and comprehensive understanding of the cosmos, the basic laws of how the universe operates, and the wonderment of science, scientific method, and inquisitive ways of thinking.

Curiosity is the primary driver for human's desire to explore the universe. In addition to teaching visitors about astronomy, the museum aims to inspire curiosity about nature and the universe, because this form of interrogation drives scientific development and invites more people to become invested in the search for scientific truth.

The starry sky is the bridge that connects humans and the universe. Humanity would have never known the existence of the universe in the absence of stars, not to mention the mysteries of how celestial bodies navigate the night skies. Regrettably, it is difficult for people in Shanghai to clearly see stars due to the bright illumination caused by intense urbanization. But on a more positive note, the Shanghai Astronomy Museum offers an opportunity for city-dwellers to see mesmerizing, crystal-clear, starry skies in their most advanced planetarium and to observe celestial bodies using its telescopes. In the core exhibition area, a variety of immersive and interactive experiences are created to allow visitors to appreciate the universe from a range of perspectives, to understand the laws of how the universe works, to learn the history and methods of how people explore the universe, and to contemplate the future of humanity.

Facing the vast and boundless universe, we humans are insignificantly minute. Yet through continuous exploration and search for scientific truths, we can try to learn and understand this grand universe. It is scientific thinking and exploration that makes humans the greatest species on Earth.

Visitors are immersed in thought-provoking exhibits while touring the Shanghai Astronomy Museum, for an entertaining journey. More important, the intention of the museum is to lead visitors to think about the ultimate questions: Where do we come from? What are we? Where are we going?

An Awe-Inspiring Journey
Main exhibition area

The museum's main exhibition area consists of three sections — "Home," "Cosmos," and "Odyssey" — which provide a panoramic view of the solar and galactic systems, how the universe operates, and how humankind explores the universe, creating a journey of multi-sensory immersion to help visitors shape a comprehensive view of the universe.

The "Home" exhibition area presents a mysterious and mesmerizing starry sky. With the help of the world's most advanced optical projectors, visitors can marvel at the Milky Way. This amazing sight, not easily seen by urbanites, marks the beginning of an immersive and awesome journey of the universe. As visitors roam around beneath the starry sky, they gradually understand more about the solar and galactic systems and start to wonder: Do other planets have water? How about volcanoes? Do they experience auroras? Standing in front of a dazzling array of celestial bodies with meteorites within easy reach, one wonders: What is our place in the cosmos?

The "Cosmos" exhibition area is designed for visitors with basic knowledge about modern astronomy and presents the laws of the universe using core concepts in five dimensions: time-space, light, elements, gravity, and life.

The Beijing Planetarium, established in 1957 and expanded in 2001, inspired a love of astronomy in China
北京天文馆成立于1957年，并于2001年正式开工扩建，激发了国内公众对天文学的热爱

提供一个与美丽星空亲密接触的机会，用最高等级的天象仪还原一个清澈宁静的星空，并带领观众使用天文望远镜观测真实的天体。而在天文馆的核心展示区域，他们打造了多种沉浸式、互动式的体验方式，来引导公众从不同的角度欣赏宇宙的魅力，理解宇宙万物运作的规律，了解人类探索宇宙的历程和方法，思索人类的明天和宇宙的未来。

面对广袤无尽的宇宙，人是那么的渺小；然而，渺小的人类却能通过长期的探索，应用科学的智慧，努力尝试去理解这个宏大的宇宙，正是这种科学的思考和探索使人类成了这个星球上最伟大的物种。

通过游览上海天文馆，你一定会学到许多以前不曾了解过的知识，但是建设者更希望的是创造一段有趣的旅程，带领每个人去探索那些人类终极的问题：我们从何处来？我们是谁？我们向何处去？

震撼体验之旅

主展区

三大主展区是上海天文馆的核心展示区域，由"家园""宇宙""征程"三大展区连贯而成，它们充分考虑了不同游客的需要，以不同的体验方式，分别展现太阳系及银河系的奥秘、宇宙的运行法则和人类探索宇宙的历程，共同形成对现代宇宙观的完整理解。

"家园"展区通过宏大的场景设计，营造出神秘而美丽的星空氛围。游客步入该展区的第一个展项就是"仰望星空"，全球最先进的光学天象仪投射出逼真的星空和壮美的银河，这将是相当多的城市人从未见过的震撼场景，沉浸式的惊诧星空之旅就此展开。游客们漫步在漫天繁星的太空中，逐渐加深对太阳系和银河系的了解：其他星球上有水吗？有极光吗？有火山吗？……琳琅满目的天外来客——陨石就在眼前触手可及，

The exhibits in this area are interactive. While enthusiasts will gain in-depth astronomical knowledge, those who are beginning their studies will get a glimpse into contemporary astronomy. They may find it difficult to understand many of the subtleties, but the exhibits will arouse their curiosity and help them learn.

The "Odyssey" exhibition area, as the name implies, offers a panoramic view of humankind's long journey to explore the mysteries of the universe, from ancient archeological digs to the first astronomical observatories to today's unmanned satellites orbiting Earth, including the Chinese Chang'e Lunar Exploration Program, space labs, and Mars exploration missions. This area relies on scientific stories and outlook on the future to inspire curiosity about the methods and spirit of scientific exploration and to encourage visitors to reflect on astronomical culture.

Four Observation Instruments

The Shanghai Astronomy Museum boasts four fascinating large-scale stargazing demonstration and observation facilities.

The first is the optical projector in the "Home" exhibition area, with an inner diameter of 17 meters. It is the soul of a traditional planetarium. The museum houses the latest projector, ORPHEUS, from GOTO INC in Japan, which offers the most realistic simulation of the starry sky.

The second apparatus is the 23-meter-diameter dome theater, which is one of the three primary forms of the museum building, a breathtaking "floating planet." Inside this structure is a set of the world's most advanced 10K-resolution dome projection systems, along with the most advanced laser and stage performance equipment. With these systems, eye-opening cosmic videos and enchanting "star concerts" are presented.

Third is the patented walk-in adaptive optical solar telescope (EAST), which allows visitors to witness a ray of sunlight turning into high-resolution multispectral images of sunspots, prominences, and flares through state-of-the-art optical systems.

The fourth apparatus is the one-meter bifocal telescope (DOT), the largest astronomical telescope in China. It can be used for both scientific observations and stargazing tours, allowing viewers to clearly see the moon, planets, and deep-sky objects at night. Even just a brief glimpse is unforgettable.

Ancillary and Educational Areas

In addition to the main exhibition area, the Shanghai Astronomy Museum has ancillary zones that are designated for temporary thematic exhibitions.

The "Chinese Exploration of the Sky" exhibition area provides a comprehensive view of China's astronomical studies in ancient times, the history of interactions on astronomy between China and the West in modern times, and contemporary developments in Chinese astronomy. The "Curious Planet" zone has a playground, dedicated to preschoolers, with the theme of adventures on an extrasolar planet. The "Sailing to Mars" zone is furnished as a Mars base, giving tourists an opportunity to appreciate the galactic mysteries of the red planet.

Additionally, the museum hosts the Star Exploration Camp, comprising an astronomy lab, a meteor lab, and a maker lab, where students can conduct experiments, listen to gripping science stories, and debate topics of interest. Those who are lucky may even camp in this zone and see the stars overnight!

In the infinite realm of unknowns, science motivates us to passionately ponder. That is where its charm and power lies. We hope your tour of the brilliantly designed Shanghai Astronomy Museum will be an entertaining journey to explore the vastness of our universe.

令人不禁好奇：我们究竟身处宇宙的何方？

"宇宙"展区面向较高的学习层次，采用了独特的主题式设计，分为"时空""光""元素""引力"和"生命"五个部分，用现代天文学最核心的几个概念，从不同的侧面来展现宇宙的运行规律。这里的展品更注重互动性，爱好者们可以学到更多高深的知识，普通游客则可以便捷直观地了解当代天文学的全貌。你可能一时难以理解其中的许多奥妙，但是你的好奇心一定会被激发，你的宇宙观也在不知不觉中得到了充实。

"征程"展区，顾名思义就是全景式展现人类探索宇宙奥秘的漫漫征途，从古人的思考到近代天文学革命，再到当今世界各地的天文台及研究计划、各种天文卫星，更有中国人自己的嫦娥探月、空间实验室和火星探测计划。这个展区更多地依托科学故事和对未来的畅想，来激发游客对科学探索方法和精神的求知欲，以及更深层次的文化思考。

四大重器

上海天文馆拥有四个特别吸引人的大型星空演示和观测设备，值得期待。

第一个重要设备就是前文介绍过的"家园"展区中的光学天象仪，内径17米，这也是传统天文馆的经典灵魂。天文馆引进日本五藤光学研究所最新型的ORPHEUS光学天象仪，将带来最为逼真的星空之体验。

第二个重要设备是直径23米的球幕影院，它的外形正是主建筑的"三体"之一——令人瞠目结舌的"漂浮星球"。其内部是一套全球最先进的10K分辨率球幕投影系统，配以激光表演系统和舞台表演系统，带给你震撼心灵的宇宙大片体验，还有梦幻般的"星空音乐会"等着你来欣赏。

第三个重要设备是上海天文馆拥有设计专利的"步入式自适应光学太阳望远镜"（EAST），游客可以亲眼目睹一缕阳光从天而降，经过最先进的自适应光学系统，变身为多个波段的高清晰度太阳像，诚如亲见科学"魔法"。

第四个重要设备是"双焦点可切换式一米望远镜"（DOT），它是国内最大口径的科普型天文望远镜，它既可以用于科研观测，也可以在夜间让你亲眼观赏最高清晰度的月球、行星和深空天体形象，仅需短暂一瞥，即可让你终生难忘。

辅助展区及教育区

主展区之外，天文馆还特别设计了几个富有特色的辅助展区。

"中华问天"展区将全方位地展现中国古人对星空的探索和思考、近代中西方天文交流的历程，以及中国现代天文学的发展。"好奇星球"为学龄前儿童专门营造了一个以梦幻般的系外星球探险为主题的游乐场所。"航向火星"则可以让部分幸运的游客置身于未来的火星基地，亲身体验火星的神秘世界。

此外，在"星空探索营"里，上海天文馆还建立了特色天文教室、陨石实验室和创客实验室，学生们可以在这里开展各种科学实验，聆听精彩的科学故事，讨论有趣的科学问题，幸运的学生还可能在营地里宿营，真正体验星空的奥妙。

面对无穷无尽的未知世界，敢于去思考，善于去思考，这正是科学的魅力和力量所在。希望这座精心设计的上海天文馆能给你留下遨游宇宙的美好记忆。

The universe is endless life with abundant power, meanwhile, it also carries a neat order with perfect harmony.

宇宙是无尽的生命，丰富的动力，但它同时也是严整的秩序，圆满的和谐。

— Zong Baihua, 1897—1986
宗白华 (1897—1986)

Connecting Humans to the Universe
连接人和宇宙

By Thomas Wong
托马斯·黄 撰文

Thomas Wong is the designer of the Shanghai Astronomy Museum and a partner at Ennead Architects. In his 30-year career as an architect, he has designed a broad range of buildings across typologies and geographic regions that have received global recognition.

托马斯·黄是上海天文馆的主创设计师，也是艺艾德建筑设计事务所的设计合伙人。从业30年来，托马斯·黄先生在世界各地经手并完成了诸多广受赞誉的各类建筑实践作品。

For the last twenty years, I have commuted each weekday some fifty miles from my home in Hunterdon County, New Jersey, to the Ennead "mothership" in Manhattan. While it may seem absurd to spend so much time on the road, the arrangement has enabled me to simultaneously inhabit two distinct worlds: the magnetic vibrancy and dynamic energy of New York and the bucolic reprieve and undulating hills of the countryside.

There is something magical about traversing these environments within the course of a day. Typically leaving the office late at night, I tread with downward gaze through the vibrant Meatpacking District, my mind flashing between a bit of envy for the fashionable, bar-hopping throngs and the latest design problem that is preoccupying me — musings that continue through the silence of the long drive home. An hour or so later, I am far from the bustle and absorbed in the serenity of the natural world, agape at the brilliance of the stars and the spotlight of the moon. On clear nights, the moon shines so brightly that the shadows of the trees and our house are vividly clear, like cut-out profiles laid gently across the still canvas of the ground plane.

在过去的20年里，我每个工作日都要通勤大约50英里（1英里=1.61千米），从新泽西州亨特敦县的家一路开到位于曼哈顿的艺艾德总部。在路上花费这么长时间似乎有些本末倒置，但我却同时拥有了两种截然不同的体验：既能在纽约感受活力四射的都市氛围，又能在新泽西欣赏山峦起伏的田园风光。

每天穿梭于截然不同的环境中总会让人有些光怪陆离的错觉。在深夜离开办公室时，每每瞥见楼下热闹非凡的肉库区（Meatpacking District），我的脑海里总会萌生许多念头，有时是羡慕那些穿着时髦、在酒吧里狂欢的人们，有时则是对新出现的设计问题的思考。漫长的回家路上，车内一片寂静，这些问题一直盘踞在我的脑海里。大约一个小时后，我远离了城市的喧嚣，沉浸在自然界的宁静中。抬眼望去，星月朗照，树林和家的轮廓清晰可辨，柔和地点缀在远方的地平线上。

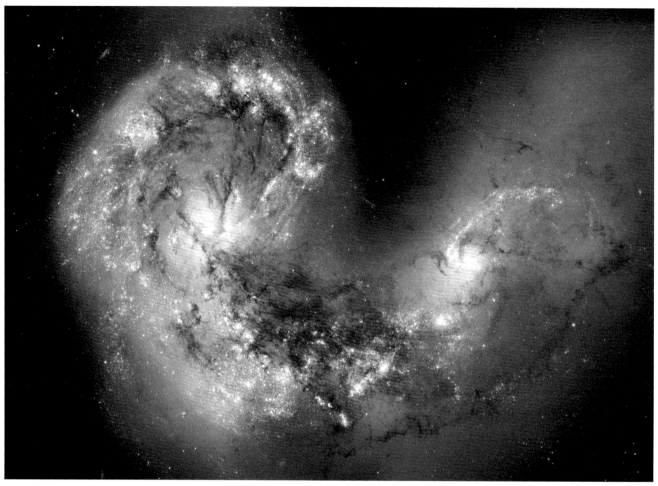

The Antennae Galaxies are a pair of interacting galaxies in the constellation Corvus
触须星系是乌鸦座中一对交互作用星系

On summer weekends, I typically suspend my life as an architect and try my hand as a novice farmer, putzing around the seven acres surrounding our 200-year-old farmhouse in my 1953 Farmall tractor. At day's end, I'll be in our back field, looking at the rise and fall of the hills that ring our valley. Given the contrast to my work week in the dense city, it's a powerful experience for me to stand within the open landscape, with unobstructed and unlimited contact to the dome of sky, the hillside cutting off the horizon in the relatively close foreground, the few adjacent neighbors and fields consumed by the expanse above. My attention is drawn upward to this vast, deeply mysterious realm, which too often goes ignored.

在夏天的周末,我通常会暂时放下工作,尝试做一个新手农民,开着我那台1953年的Farmall拖拉机,在有200年历史的农舍周围7英亩(约2.8公顷)的土地上晃悠。一天结束时,我会站在农舍后面静静凝望连绵起伏的山峦。在高楼林立的城市工作一周后,站在这幽幽山谷之中,顿感心旷神怡。纵目四望,天地辽阔,山坡上星星点点散落着几户人家。这片天地的广袤与神秘深深地吸引着我,可它的美丽却鲜为人知。

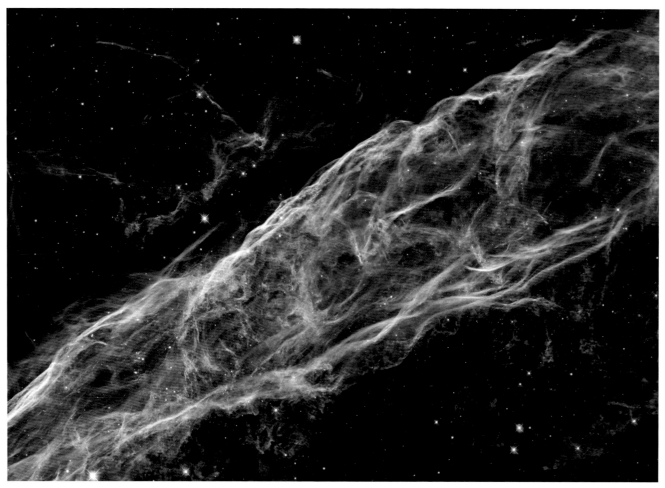

The Veil Nebula, in the constellation Cygnus
天鹅座面纱星云

In thinking about the Shanghai Astronomy Museum and that powerful connection to the sky, I am struck by a realization about my daily commute: That same fifty-mile trip, if taken in the vertical direction, would bring me to a place far more transformative than the difference between my city and country environments. Some fifty miles upward is where the physics of flight fundamentally change, as the atmosphere becomes so thin that the aerodynamics of lift via pressure differential no longer apply — a place where there is a dissipation of atmospheric drag, and space-based flight prevails. This threshold, known as the Kármán Line, is by many definitions "the edge of space," the end of the Earth-bound and the beginning of the infinite expanse, the point of departure from our planet's atmosphere. Amazingly, this place lies close enough for what I consider to be routine travel.

在思考上海天文馆的设计以及与天空的强大联系时，我对我的日常通勤突然有了一种新的认识：如果沿着垂直向上方向同样走50英里，我将来到一个全新的天地，环境变化的剧烈程度将远远超出城乡之别。在那里，因为大气层非常稀薄，依靠压力差抬升高度的空气动力学不再适用——大气阻力消散，物理条件发生了根本性变化，飞行以太空飞行为主。这个阈值被称为卡门线（the Kármán Line），它被定义为"太空的边缘"，是外太空与地球大气层的分界线。令人惊讶的是，这虽是地球的尽头，但却又仿佛和我日常通勤所经历的一切如此相似和接近。

The Bubble Nebula, in the constellation Cassiopeia
仙后座的泡泡星云

Of course, it is a far more complex endeavor to move in that upward trajectory than it is to journey down Interstate 78 to the Holland Tunnel. But the proximity of the Kármán Line shines light on a fairly astounding but basic truth: Humans (and indeed, the vast majority of life on Earth) exist within an infinitesimally thin slice of physical reality. Within that shallow plane — its thickness relative to the distance from my driveway to the highway entrance ramp — resides the entire human construct, the totality of our history, our accomplishments and devastations, the ridiculously small realm in which our lives individually and collectively play out. Inside of this micro-plane of atmosphere that envelops our increasingly fragile planet, we live in a bubble that allows for a stunningly myopic existence, one in which the infinite universe seems far away and considerably (though mistakenly) less compelling than whatever is flashing on our smartphones.

One of the primary goals that informed the design of the Shanghai Astronomy Museum was the need to expand the modern human perspective, to create an experience that strengthens our awareness of the realm beyond our planet and illuminates our place in the universe. We wanted to create a space that would make evident the astronomical truths that make our very existence possible and help us understand how exceptional Earth's life-supporting aspects are when compared to the turbulent environments elsewhere in the galaxy, and beyond.

当然，沿着这条向上的轨迹前进要比沿着78号州际公路（Interstate 78）到达荷兰隧道（Holland Tunnel）复杂困难得多。但是卡门线的临近，揭示了一个相当惊人但又基本的事实：人类（以及地球上绝大多数生命）存在于一片极薄的实体中。实体之薄，就像眼下的车道到高速公路入口匝道这么近。但就在这样一个极薄的、极狭小的层次内，包含着人类全部的造物、全部的历史、一切成就与破坏，上演着芸芸众生的生老病死。这个超薄的大气层包裹着我们日益脆弱的星球，人类就像生活在一个气泡中，变得极其短视，沉迷于眼前的智能手机，而遗忘了遥远宇宙的无限魅力。

上海天文馆设计的主要目的之一是拓宽当代人的视野，使人们有机会去认识地球以外的领域，并了解自身在宇宙中的位置。我们想创造一个空间，清楚地向大家展示那些使人类的存在成为可能的天文奥秘，并帮助人们了解银河系其他地方以及更遥远的宇宙，从而意识到孕育着生命的地球是多么独一无二。

Jupiter, photographed by NASA's Hubble Space Telescope
木星，由美国宇航局哈勃太空望远镜拍摄

Civilizations of the past utilized the built environment to achieve just this; far more attuned to the astronomical phenomena that shaped their existence, they built structures to mediate a fundamental relationship that may not have been scientifically understood but was intrinsically valued. The near-perfect cardinal alignments of the pyramids of Giza; the orientation of the major axes in the city of Teotihuacan, which conform to an ancient concept of a solar calendar; the marking of the solstices and the path of the moon at Stonehenge; the multilayered geometric symbolism at the Temple of Heaven, which metaphorically connects Earth and sky — these architectural feats are a testament to the fascination with the universe in the collective consciousness.

Modern life has detached most of us from that elemental bond. We take for granted the functional aspects of the Earth's daily rotation, the circumscribing watch of the moon, our revolution around the sun — a reality that usually becomes filtered through a digital clock on an app on a tiny screen. Situated within this contemporary context, the institutional mission of the museum and the architectural concept of the building are aligned: to engage our minds and spirits in an experience that ignites curiosity, transports us beyond the quotidian, and inspires exploration of the Great Beyond.

在古代，人们利用建成环境来实现这一目的。古人笃信天文现象与人类世界之间存在神秘的关联——当今的科学尚无法破解这些联系——并深深崇敬着它们，于是建造了许多构筑物，试图与宇宙进行沟通：例如吉萨金字塔群所蕴含的近乎完美的数字巧合；特奥蒂瓦坎城主轴的方向与古代太阳历所记录的重要方位一致；英格兰巨石阵中二至点和月亮轨迹的标记；北京天坛的多层几何形式隐喻了天空与大地的联系……这些建筑成就是集体意识对宇宙美好想象的证明。

现代生活已经使绝大部分人脱离了这种纽带关系。昼夜的交替、月亮的盈亏、四季的更迭都被简化为手机屏幕或时钟程序上跳动的数字，再无法引起我们深究的兴趣。在这样的背景下，天文馆的使命和建筑物的设计概念殊途同归了——通过激发好奇心来调动人们的思想和精神，引导他们突破日常，去探索更宏大的未知。

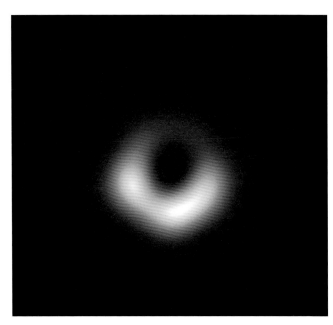

A black hole in galaxy Messier 87. The black hole is 6.5 billion times the mass of the Sun

M87星系中心黑洞，其质量是太阳的65亿倍

The museum is meant to inspire more questions than it answers. What is the nature of places that are "out there," so incomprehensibly far from Earth and unimaginably strange, yet governed by many of the same physical laws of science? Could one of these places host life? What technology have we developed to make them real and comprehensible? Eventually, such questions pivot toward self-reflection: What are we — humans and all other life on our planet — in relation to everything else? What is the role of our relatively microscopic reality within the universe? What are the risks to Earth's delicate ecosystems, and how must we respond?

Architecture of and for the Cosmos

We made this building in the hope of sharing excitement about the known and yet-to-be-discovered wonders of space, as a place where the educational mission of science is enmeshed with an architecture, born of deep conceptual origins, that itself is also teaching.

A foundational design concept was to shape the architecture around those magnificent elements dynamically alive among the stars, to abstractly embody the phenomena and laws of astrophysics that are the rule in space. A powerful spatial experience would serve to reinforce the monumentality and grandeur of the museum's subject matter and would heighten visitors' attention and awe to forces beyond imagination. To create an architecture that speaks to a realm "beyond" meant to craft a spatial and experiential language that originates from interstellar

因此，天文馆的存在是为了激发思考，而非提供答案。那些"远在天边"的地方到底是什么？它们距离地球如此之远，于我们而言又是那么难以想象的陌生，是否也受自然物理法则的支配？其中的某个地方也存在生命吗？为了证实和理解这些未知的地方，我们发展了哪些科学技术？这些问题最终都导向了自我反思：我们——人类和所有生存在地球上的生物——与其他事物的联系是什么？看起来微不足道的我们在宇宙中又扮演着什么样的角色？脆弱的地球生态系统面临着什么威胁？我们又必须如何应对？

宇宙的构筑

上海天文馆是为了分享探索宇宙空间的激情而建，除原定的科普功能外，我们也希望在建筑本身的设计中赋予深意，使之寓教于形，经得起推敲。

Venus at sunrise, as seen from the International Space Station
从国际空间站看到日出时的金星

matters. One fundamental notion became a primary source of invention: the fact that the universe, from the time of the Big Bang, is in a state of perpetual motion. From the expansion of galaxies over billions of years to the complex gravitational relationships of astronomical entities acting upon one another, the building design originates from the dynamic energy of celestial movement.

Lunar orbits around their respective planets, as well as the orbital systems surrounding individual stars, are among the most fundamental frameworks of organization in the cosmos. Our own solar system spins amidst the greater spiraling force of the Milky Way galaxy. The intricate choreographies created by gravitational attraction of multiple bodies yield orbital paths of incredible dynamism. (The complexity of such motion is detailed in the unsolvable classic physics model known as "the three-body problem").

一个基本的设计概念是参考那些在星空中跃动的巨大元素来塑造建筑，抽象地体现那些支配宇宙的天体物理学现象和规律。富有冲击力的空间体验可以让天文馆倍显宏伟震撼，使参观者对那种凌驾于人类想象之上的力量更感敬畏。要创造一个与"遥远之地"进行对话的建筑，意味着要创造一种源于星际的体验式空间语言。这种创造来源于天体运动的动态能量：从宇宙大爆炸开始，宇宙就一直处于这种动态之中——从数十亿年来星系的膨胀到天体之间的复杂引力关系莫不如此。

恒星及其行星绕行系统，以及这些行星的卫星绕行系统是宇宙中最基本的组织架构之一。我们所处的太阳系则在银河系更大的作用力下转动。多种天体之间的引力作用错综复杂，产生了不可思议的动态轨迹（这种运动的复杂性在无法解决的经典物理学模型"三体问题"中有详细说明）。受此启发，天文

This inspired the curving architectural ribbons of the museum's facade. The building's envelope traces a series of arcing paths that are visibly influenced by gravitational pull: the heart of the central atrium, the forward momentum at the entry, the planet-like sphere that envelopes the Dome Theater.

Building upon the concept of dynamic orbits, feats of "structural magic" were employed to perform architectural illusions that simulate weightlessness amidst the vacuum of space. As a culmination to the orbital motion of the massing, the structure of the building defies the gravitational weight of the Earth, its vaulting uplift at the entry cantilever in counterpoint to the rooted elements of the other half of the building perimeter. The museum gestures toward the sky and space, signaling the ensuing journey of exploration within.

Likewise, the planetarium sphere appears suspended within the larger museum massing with little visible support — a veritable moon that is escaping the Earth's gravity. The pure spherical form references the primordial shapes in our universe and, like the orientation we yield from our position relative to the sun or moon, becomes an ever-present reference point for the visitor. Embedded in the roof plane of the lower museum wing, as if rising out of the Earth-bound horizon, the sphere gradually emerges into view as one rounds the building, the drama unfolding as though one were approaching a planet from one of its moons. These perceptual experiences metaphorically transport the visitor to another realm, abstractly embodying the visual sequences one might encounter beyond Earth.

The manifestation of dynamic motion continues inside the museum, where a 720-degree spiraling ramp animates the central atrium. The ramp is a centerpiece that traces the orbital flow of the visitor sequence throughout the museum exhibits and launches the eye upward, toward the oculus of the inverted dome. A heroically monumental concrete tripod cradles both the inverted dome and the spiraling ramp; the elements assemble into a central set piece, serving almost as an interstellar baldachin.

The typological reference to places of worship is intentional: Domes, altars, a nave, an oculus, ambulatories and apses — all can be found in the building. Like the great cathedrals of the West, or ancient sanctuaries such as the Pantheon or the Temple of Heaven, the building aims to connect humans with a mystery and power that is beyond Earth, becoming a metaphysical conduit.

A Global Perspective

The Shanghai Astronomy Museum fits squarely within the paradigm of Ennead's work, as it shapes the surroundings well beyond the footprint of a building, extending far into the public realm. Through this design

馆外立面修建成了弯曲的条带状，其外围形态串联起一系列喻示着天体引力影响的曲线路径——从中央中庭的中心、入口处的前奏景观，以及内部形如行星的天象厅球体，皆是天体间万有引力的一种具象表现。

在动态轨道概念的基础上，我们采用"结构魔法"来构造建筑，模拟宇宙真空中的失重状态。作为轨道运动的高潮段落，天文馆的建筑结构以入口悬臂处的超高隆起与建筑外围另一侧稳坐于地面的元素形成对立，暗喻着对地球重力的抵抗。整个建筑呈现出拥抱天空的姿态，标志着接下来的内部探索之旅。

同样，天象厅球体看起来像是悬浮在天文馆中，几乎看不见支撑物，仿佛正在逃离地球引力束缚的月球。其正球体的形状参考了我们宇宙中的原始形态，就像我们根据自身相对于太阳或月亮的位置来确定方向一样，这个球体对来访者来说也是一个永恒的参照点。它镶嵌在天文馆下翼的屋顶上，仿佛从地平线上升起，当人们沿着建筑绕行时，球体逐渐出现在视野中，形成富有戏剧性的一幕，仿佛乘着一颗卫星去接近它的行星。这种感知体验使参观者在不知不觉中进入另一个抽象空间，去面对地球以外的景象。

在天文馆内部，动态的概念继续得以展现。在馆内中庭，一个720°螺旋形坡道作为核心体量，掌控了整个观展空间的动势，它引导着人流如轨道上的群星般移动，并将其视线不断引向上方，直至触及顶端的倒转穹顶。壮观的混凝土三脚结构支撑着倒转穹顶和螺旋坡道，这些元素组合在一起构成了天文馆的中央场景，如同一个星空的华盖。

天文馆的建筑设计也有意参考了中外祭祀建筑的多种空间类型，如穹顶、中庭、圆洞天窗、回廊、半圆室。就像教堂、万神庙、天坛祈年殿这样的古代圣殿一样，这座建筑旨在将人类与遥远天外的神秘力量联系起来，成为一个超自然的连接。

philosophy, we seek to utilize architecture to strengthen the civic institutions of our age, to manifest the collective values of our society, and to address subjects of profound importance to our futures.

The museum is meant to be one of a network of institutions that are repositories for the colossus of human intellect, promoters of understanding around the globe, inspirers of new research and discovery.

Space is a great unifier, and perhaps astronomy is the paragon of exemplary cooperative study; astronomical research is conducted through close international collaboration by the collective use of exploratory tools such as the Gemini Observatory in Hawaii, the South Pole Telescope in Antarctica and the Atacama Large Millimeter Array in Chile or through transnational ventures such as the International Space Station. Space exploration brings humanity together, dissolves our fabricated boundaries and promotes the sharing of scientific ideas, theories, and knowledge, as well as the collective mission of human endeavor.

But our hope is that the reach and impact of this museum stretch beyond this. I imagine that a thousand years from now, human civilization (if it exists!) may look at the Shanghai Astronomy Museum as a structure that embodied the values of an era, spoke to the forward aspirations of humans, and connected people around the planet to the limitless expanse beyond.

The museum is a place where I hope visitors will be reminded of an important universal perspective: where we humans sit is in relation to the existence of all things, both near and impossibly distant. It is a place where I hope we will acknowledge the great fortune of Earth amid the unimaginable hostilities of the cosmos in tandem with the underlying responsibility to care for this planet and all species of life here.

My greatest aspiration for this museum is to stir thoughts beyond the human construct, to encourage us to transcend our divisions and obsessions, and to examine ourselves individually and collectively in light of the fact that we are all together in a fragile, irreplaceable lifeboat floating in the limitless void of space: the Pale Blue Dot. This is an invitation to shift away from a myopic viewpoint and discover the infinite universe that lies beyond us and the wonders in need of our care that are right in front of us.

Thomas Wong
New York City and Lebanon, New Jersey
January 2021

全球化视野

自1963年成立以来，艺艾德始终致力于创造具有公共影响力的建筑作品，重塑公共建筑与社会以及人的关系。这一点在上海天文馆的建筑设计上得到了淋漓尽致的展现。基于这样的设计理念，我们希望用建筑来加强公民设施在这个时代的作用，体现社会集体价值观，并回应一些对未来具有深远意义的问题。

上海天文馆是一座公民设施，更是人类巨大智慧的储藏库。它意在推动人们对地外空间的认知，为新研究和新发现提供灵感来源。

太空是一个宏大的整体，天文学也是合作研究的典范。天文研究的开展有赖于紧密的国际合作，以及太空探索工具的共享，如夏威夷的双子座天文台、南极洲的南极望远镜、智利的阿塔卡玛大型毫米波天线阵，以及国际空间站等。太空探索将人类团结在一起，消除了人为造就的壁垒，使科学思想、理论和知识得以共享，服务于人类不断开拓进取的共同使命。

但我们希望上海天文馆的影响力远不止于此。或许千年之后，它能在人类文明（如果那时人类依然存在的话）历史上留下一笔，成为一个体现时代价值、表达人类愿景，并将人类与无垠宇宙连接起来的里程碑。同时，它能向所有参观者传达这样的宇宙观：无论是近在咫尺还是无限遥远的存在，都与人类有所关联。在充满未知的宇宙中，地球是我们能够掌控的巨大财富，我们要保护好这个星球以及星球上的所有物种。

从人工造物的桎梏中挣脱、摒弃分歧和执念；珍惜地球这艘无垠星海之中独一无二的生命之舟，正视它的脆弱，也正视人类作为个体和集体之于它的关系——这是我寄予上海天文馆这座建筑最大的愿望。现在，就让我们打开视野，去探索地球之外的无限宇宙，以及那些近在眼前、亟待关注的奇观。

托马斯·黄
纽约州纽约市及新泽西州黎巴嫩镇
2021年1月

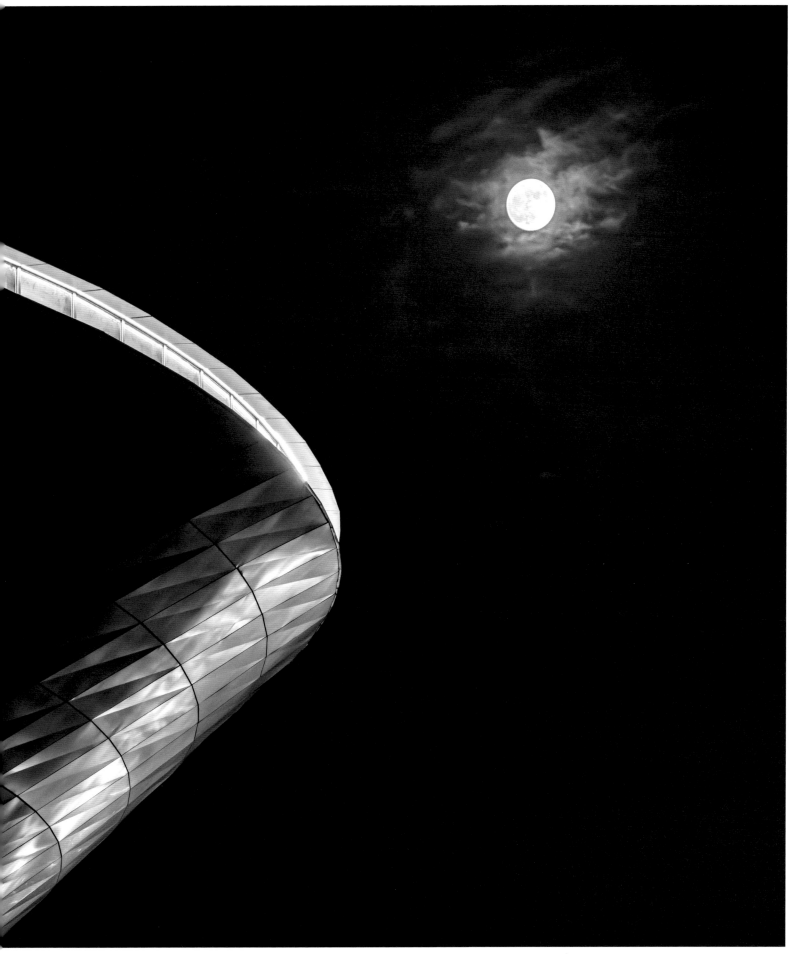

Site
场地

The Shanghai Astronomy Museum is a place where we hope visitors
will be reminded of a shared universal perspective:
where we humans sit in relation to the existence of all things both
near and impossibly distant.

我希望上海天文馆能向所有参观者传达这样的宇宙观：
无论是近在咫尺还是无限遥远的存在，都与人类有所关联。

— Thomas Wong
托马斯·黄

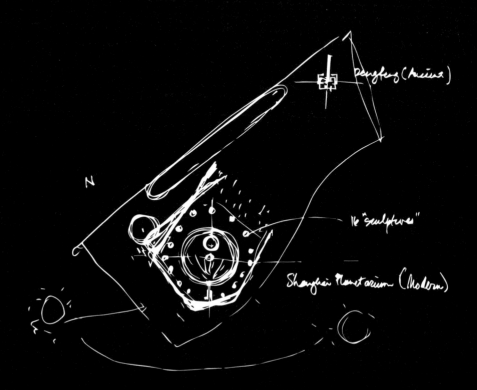

View to the east, with the round Dishui Lake and the East China Sea beyond
向天文馆东侧眺望，可以看到圆形的滴水湖，以及更远处的东海

Shanghai 1984
Population 6,828,000

1984年的上海
人口数： 6 828 000

The Shanghai Astronomy Museum is located in Lingang, a hypermodern sub-city about 70 km southeast of downtown Shanghai.

Built on land reclaimed from the East China Sea, Lingang was conceived as a harbor center to address the explosive growth of Shanghai. The world's third-largest city (after Tokyo and Delhi), Shanghai had a population in 2018 of more than 24 million, according to the United Nations, and is expected to grow to nearly 30 million by 2030. Lingang — one of a handful of "new towns" planned by the Chinese government to create structure and accommodate the needs of the megalopolis — comprises an industrial zone, recreation areas and residences for 830,000 people. The area is accessible by the Shanghai Metro system; under development is a line connecting Lingang directly to Pudong International Airport, expected to be a 15-minute ride.

For citizens residing in the heart of Shanghai, where the urban glow makes for difficult viewing of the night sky, or visitors to Shanghai Disney nearby, an easy trip to the museum on the outskirts of the city offers improved conditions for celestial observation.

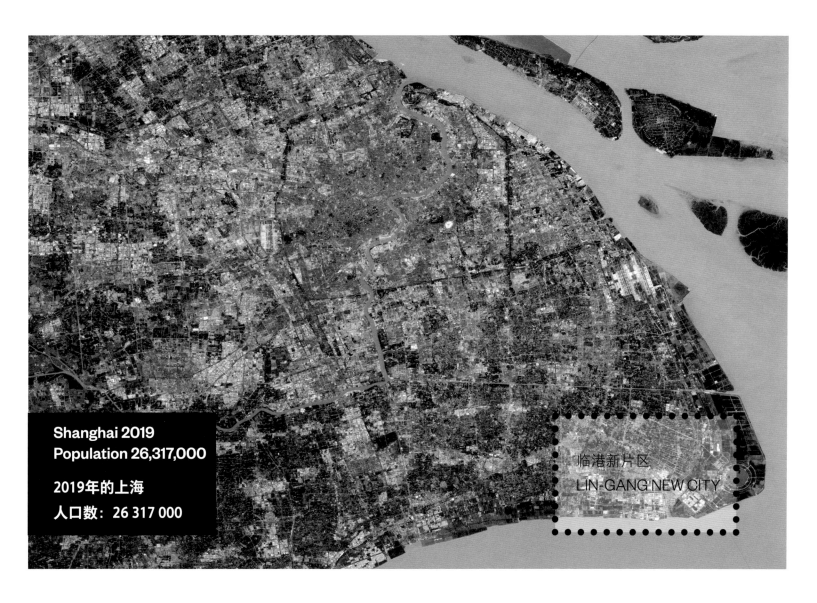

Shanghai 2019
Population 26,317,000

2019年的上海
人口数：26 317 000

临港新片区
LIN-GANG NEW CITY

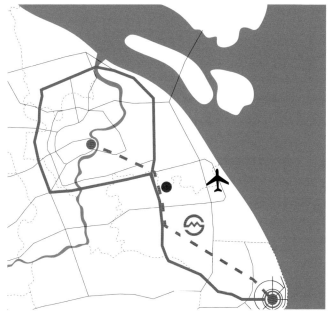

上海天文馆坐落于距离市中心约70千米的临港新城滴水湖畔。

临港新城建在东海沿岸填海而造的土地上，计划将其打造为港口中心，以应对上海人口的爆炸性增长。根据联合国的数据，上海是世界第三大城市（仅次于日本东京和印度德里）。2018年，上海人口数为2423.78万，预计2030年将增长至近3000万。正在进行建设的地铁线路将使从浦东机场到临港新城的通勤时间缩减至约15分钟。

在闪耀的霓虹灯下，居住在市中心的市民很难看到夜空。对于这些市民和上海迪士尼附近的游客来说，前往市郊天文馆的交通十分便捷，这也为人们进行天体观测提供了更好的条件。

Lingang is built around a manmade, circular body of water, Dishui Lake, with a diameter of 2.5 km. Streets are arranged around the lake in concentric circles. The Shanghai Astronomy Museum creates a landmark structure and civic hub on the outermost band, defined in the Lingang master plan as a district for cultural institutions and parks.

Inspired by galactic structure, the siting of the museum suggests that the building has been attracted by Dishui Lake's "gravitational pull" and is "revolving" around it. Underscoring this concept, the relative scale of the astronomy museum's Sphere to Dishui Lake with the Huanhu Road system is approximately proportional to that of the Earth relative to the Sun.

临港是围绕着一个人造圆形水域——滴水湖而建，滴水湖直径2.5千米。街道绕湖呈同心圆布局，上海天文馆即位于其最外围的区域，既是地标性建筑，也是一个市民活动中心，该区域在临港总体规划中被定义为文化机构和公园区。

受银河系结构的启发，天文馆的选址寓意着该建筑被滴水湖的"引力"吸引，并围绕着它"旋转"。为了强调这一概念，天文馆球体与滴水湖加环湖路的大小比例接近地球与太阳的大小比例。

The diameter of the sun (1.39 million km) is 109 times greater than the diameter of Earth (12,756 km). The diameter of Dishui Lake and the middle ring road (3,161 m) is 109 times greater than the diameter of the Shanghai Astronomy Museum's Sphere (29 m).

太阳直径为139万千米,比地球直径12 756千米大109倍。滴水湖加环湖路的圆形直径为3169米,比天文馆球体直径29米大109倍。

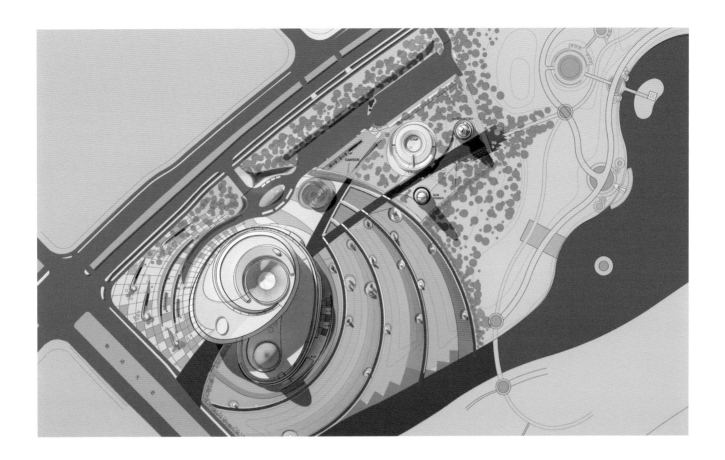

The site and landscape were designed to enforce the notion that the museum "landed" on the site, creating a tree-less impact crater of concentric arcs that emanate from the building's massing, evoking radial energy as they dissipate into the surroundings. A corten steel site wall establishes the threshold between the point of "impact" and the natural condition.

The museum's influence extends to adjacent zones of the green belt, including an adjacent 540,000-square-meter astronomy-themed public park that spans the canal.

场地和景观的设计是为了强化天文馆"降落"在此的概念，建筑师在建筑体量周围塑造了一系列无树木覆盖、呈同心圆弧状向外辐射的撞击坑，以模拟陨石落地时激起的能量波。而建筑主体的钢结构外墙在"撞击点"和自然环境之间建立起了一道屏障。

上述场地设计概念的范围一直延伸至邻近区域的绿化带，其中包括附近一座跨越河道、面积达54万平方米的"星空之境"天文主题公园。

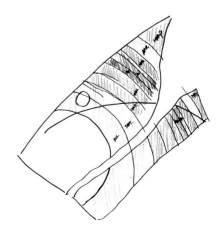

The Campus
场馆

1

1. Main Museum
2. Youth Observation Camp
3. Solar Telescope
4. Optical Telescope

1 主场馆
2 青少年天文观测基地
3 羲和太阳塔
4 望舒天文台

Auxiliary structures for the visitor experience include facilities for a Youth Observation Camp and education and research center that contains astronomy classrooms, meteorite laboratories and maker laboratories. A 23-meter-tall walk-in adaptive optics solar telescope allows visitors to observe high-definition images of sunspots and solar flares. A double-focus, 1-meter-aperture optical telescope, the largest in China, is available for use by visitors.

服务于游客的附属建筑包括青少年天文观测基地和教育研究中心，后者开设有天文教室、陨石实验室和创客实验室。游客可以用一个23米高的步入式自适应光学太阳望远镜观察到太阳黑子和太阳耀斑的高清图像。这里还有中国最大的双焦点可切换式1米口径光学望远镜。

Optical Telescope　望舒天文台

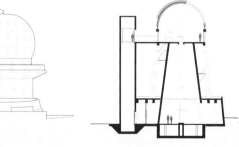 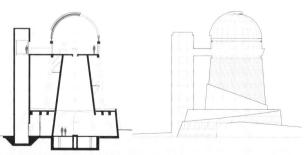

Solar Telescope　羲和太阳塔

Solar Telescope　羲和太阳塔

Optical Telescope　望舒天文台

Concept
概念

A foundational design concept was to shape the architecture around those magnificent elements dynamically alive among the stars, to abstractly embody the phenomena and laws of astrophysics that are the rule in space.

一个基本的设计概念是参考那些在星空中跃动的巨大元素来塑造建筑，抽象地体现那些支配宇宙的天体物理学现象和规律。

— Thomas Wong
托马斯·黄

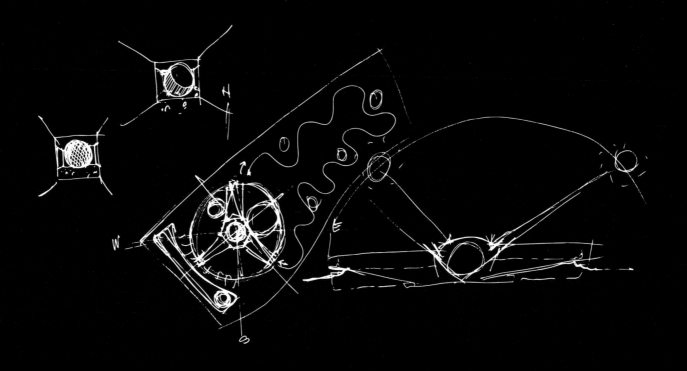

Competition model　竞赛模型

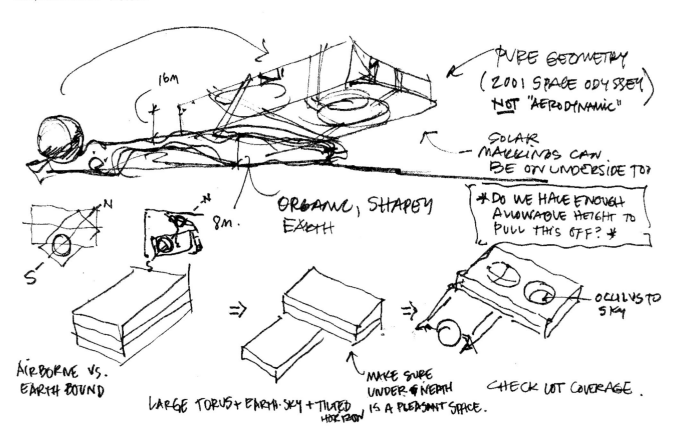

The Shanghai Science and Technology Museum, which opened in 2001, sought to enhance its exhibitions on biology, earth science, robotics, aerospace and other fields with a world-class planetarium and astronomy facility. The museum held an international competition for designs.

Ennead Architects embarked on a range of design approaches. One of six teams invited to the competition, while each was presented as a cohesive concept, the intention was to merge the most successful aspects into a single unified proposal for a rich, powerful work of architecture that would become the firm's final submission to the competition.

In 2014, Ennead's submission was declared the winner of the competition. The firm honed the design over the following eight years, working in conjunction with the Shanghai Institute of Architectural Design and Research. Groundbreaking on the Shanghai Astronomy Museum took place in November 8, 2016.

上海科技馆于2001年开馆，期望通过建设世界一流的天文馆并配备高端天文设施，提升生物学、地球科学、机器人、航空航天等领域的展览品质。为此，科技馆举办了一次国际设计竞赛。

总部位于纽约的艺艾德建筑设计事务所团队进行了一系列设计探索。作为上海天文馆建筑设计国际竞赛的六家受邀竞标单位之一，团队最终提交的设计方案内容丰富，具有极强的感染力，是由大量具有独立完整理念的初期方案经过不断融合、提炼、升华后形成的。

2014年，艺艾德有幸摘得此次国际竞赛桂冠，成为上海天文馆建筑设计的主创设计单位。在接下来长达8年的时间里，艺艾德在当地建筑设计院（上海建筑设计研究院）的配合之下对整个设计方案进行了深化、调整和完善。上海天文馆于2016年11月8日正式破土动工。

Design Exploration
设计探索

The design process was an investigation into a wide variety of architectural ideas inspired by the subject of astronomy. Here are some of the primary concepts explored in early studies, which were synthesized into the final competition proposal.

在天文学主题的启发下,我们在设计过程中对各类建筑创意进行了充分探索。以下是设计初期探索的一些主要概念,它们已被整合到最终的竞赛方案中。

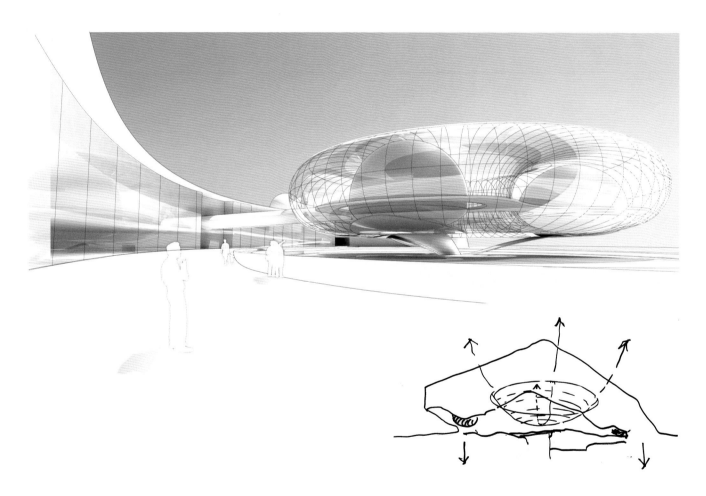

One principle design concept was to create a dialogue between the grounded and the skyward. Like many structures built by ancient civilizations in an effort to understand the world and the heavens, the museum's fundamental mission was to connect the Earth to the sky. Amid a variety of formal explorations in early schemes was a consistent goal of linking grounded elements with those reaching upward or levitating—a concept that carried through to the final design.

设计的主要概念之一是在地面和天空之间建立对话。就像古人为了解世界和天空而建造的许多建筑一样,天文馆最重要的使命是沟通地球和太空。在早期的各种形式研讨中,我们始终坚持一个理念,即把地面元素与向上延伸或悬浮的元素联系起来——这个概念一直延续到最终方案中。

Temple of Heaven (Hall of Prayer for Good Harvest), Beijing, China
中国北京的天坛（祈年殿）

Pyramid of the Sun, Teotihuacan, Mexico
墨西哥特奥蒂瓦坎的太阳金字塔

The Shanghai Astronomy Museum's architectural instruments that track the movement of the sun have functional antecedents in ancient structures that mark the passage of time and display cosmological symbolism, such as the Pyramid of the Sun at Teotihuacan in Mexico and the Temple of Heaven in Beijing.

上海天文馆的建筑空间经过巧妙设计，可以追踪太阳的轨迹，类似做法在墨西哥特奥蒂瓦坎的太阳金字塔和北京的天坛祈年殿等古建筑中也有运用。这些建筑记录了时间的流逝，也展示了宇宙的象征意义。

The Horizon

Ennead had previously examined spherical architecture in its design for the Rose Center for Earth and Space at the American Museum of Natural History in New York, arriving at an iconic sphere within a glass cube. In Shanghai, the Dome Theater program presented an obvious opportunity for the architects to revisit the sphere, a strong reference to celestial entities. In this case, the concept was for a sphere that crosses a metaphorical horizon, accentuating the visual relationship between objects in the foreground and those in the far distance.

天际线

艺艾德此前在美国自然历史天文馆的罗斯地球与太空中心设计中曾采用过球体建筑形式，并最终为公众呈现了极为经典的玻璃立方体内嵌球体的建筑形象。天文馆球形建筑和宇宙天体的自然形态有着极为默契的联系，而上海天文馆的球幕影院即是建筑师对球形建筑的一次崭新尝试。其设计概念是塑造一个跃出"地平线"的球体，以强调前景物体和远处物体之间的视觉关系。

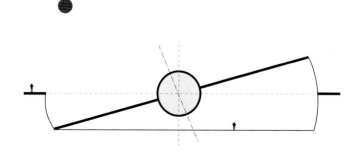

Ecliptic Plane (The imaginary plane containing the Earth's orbit around the Sun)
黄道面（地球绕日轨道所在的平面）

Equatorial Plane (The plane on which the earth's equator lies, which refers to the plane passing through the geocentre and perpendicular to the earth's rotational axis)
赤道面（地球赤道所在的平面，也是过地心与地轴垂直的平面）

The crescent Earth rises above the Lunar horizon in a photograph taken from the Apollo 17 spacecraft
在这张由阿波罗17号宇宙飞船拍摄的照片中,地球正从月球地平线升起,其昼半球形似新月

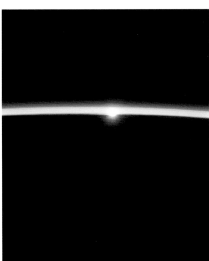

The "thin line" of the Earth's atmosphere and the setting sun
薄如一线的地球大气层,以及仿佛镶嵌其上的落日

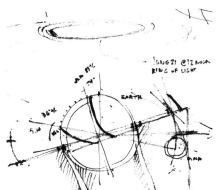

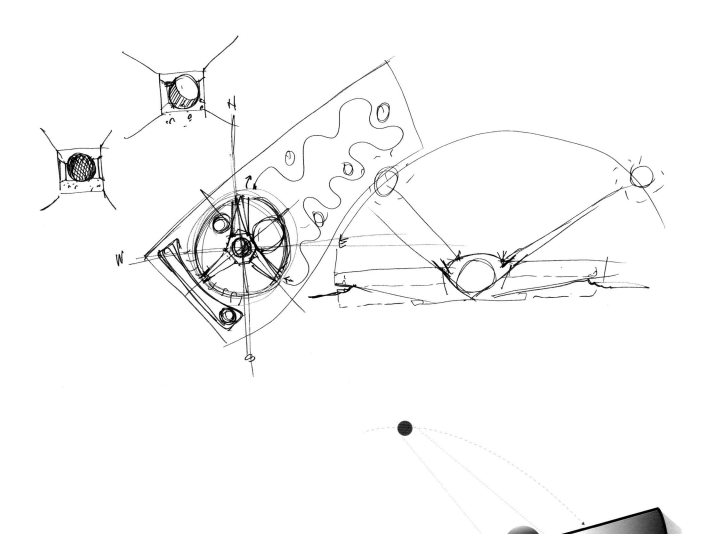

Tracking the Sun

A goal emerged early in the design process for the building to embody the fundamental astronomical principles of the Earth's rotation and its revolution around the Sun. By tuning apertures within the building to coordinate with solar positions throughout the day and over the course of a year, the architecture crafts figures out of sunlight and becomes an astronomical instrument at a monumental scale.

循迹太阳

在早期设计概念中，我们一致认为，天文馆建筑要体现出地球自转及其绕太阳公转的基本天文学原理。我们通过调整建筑的开窗，使其能够将太阳一天中以及一年中的位置变化转译为特定的光影轮廓，建筑也因此成为一种巨型天文仪器。

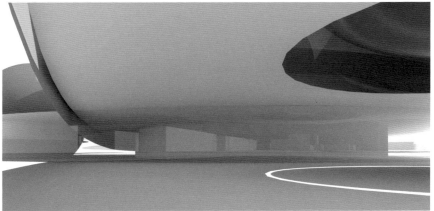

A full circle of sunlight appears beneath the Oculus each day at noon
每日正午，阳光会穿过圆洞天窗，如日晷一般，在下方投射出与地面圆形完全重合的光圈

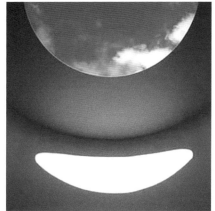

James Turrell, *Skyspace*
詹姆斯·特瑞尔，《苍穹》

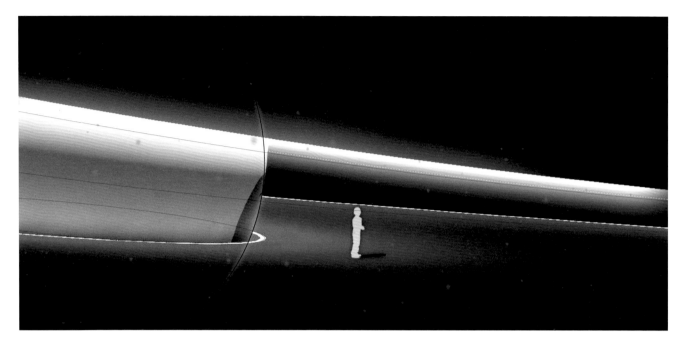

Study Models
研究模型

Spatial Experiences of Space

The premise for the interior was to transport the visitor from the urban construct to another world, unlike anything on the Earth. Through the manipulation of natural light and the creation of dramatic architectural set pieces, the visitors experience an abstract spatial journey inspired by the elements of the universe.

体验太空

馆内空间设计意在使参观者置身于一个与地球截然不同的世界。我们通过对表面自然光的巧妙利用，使参观者可以捕捉到不同寻常的瞬间，体验一场抽象美的空间之旅。

The Sphere (which holds the Dome Theater)
天象厅球体（球幕影院所在）

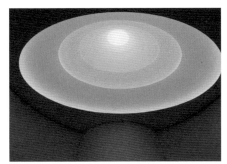

Aten Reign by the artist James Turrell, Image for image
詹姆斯·特瑞尔的艺术装置作品《日冕》，画中画

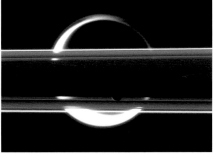

Saturn's rings seen with Titan, its largest moon
土星环与其最大的卫星泰坦

A flight tube at Eero Saarinen's TWA Terminal in New York
埃罗·沙里宁设计的纽约环球航空公司航站楼的登机通道

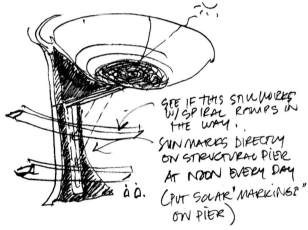

Contact with the Sky

A core strategy for the design was enabling visitors to have an experience of direct contact with the heavens by inhabiting a space that cuts the view and presence of the adjacent urban context, thereby turning attention upward with an unimpeded view of the sky. This very real experience is the culmination of the visitor journey and complements the virtual and simulated nature of the universe in the exhibit halls.

对话苍穹

设计的核心策略之一是让参观者能够与天空直接对话——在一个与周围城市环境相隔绝的空间里仰望天空，尽情徜徉在浩瀚无边的宇宙世界里。这种极度真实的体验是参观过程中的高潮，也对展厅中模拟的宇宙规律进行了补充。

倒转穹顶的早期渲染图

光与空间艺术家詹姆斯·特瑞尔的作品
《罗登陨石坑》

夜间长曝光拍摄的星空

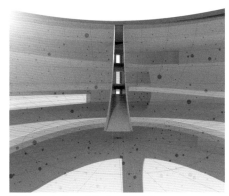

Early rendering of the Inverted Dome
倒转穹顶的早期渲染图

Roden Crater by light-and-space artist James Turrell
光与空间艺术家詹姆斯·特瑞尔的作品
《罗登陨石坑》

Long exposure of the sky at night
夜间长曝光拍摄的星空

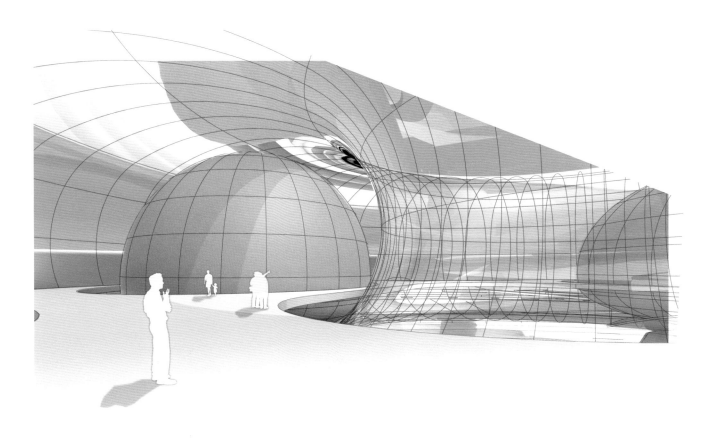

Dynamic Orbital Motion

Drawing inspiration from the fact that the universe is in constant motion, the museum references dynamic orbits in its forms, sequences and gestures. Mimicking the gravitational pull of interstellar matter and the complex dance of moons orbiting planets and planets orbiting stars, the building espouses energy as its massing suggests a constantly spinning momentum.

动态轨迹

宇宙是永动的。天文馆从中汲取灵感,在建筑形式、序列设计和外观上均参考并模拟了星际物质之间的引力,以及卫星围绕行星、行星围绕恒星的错综复杂的运动轨迹,仿佛将驱动天体旋转的能量蕴于自身之中。

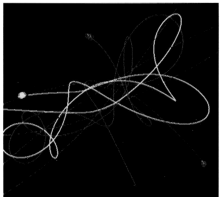

An illustration of the chaos depicted in the "three-body problem"

一幅描绘"三体问题"中混沌状态的插图

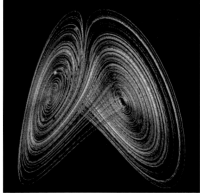

A diagram of the Feigenbaum constants

费根鲍姆常数图示

Early rendering of Inverted Dome

倒转穹顶的早期渲染图

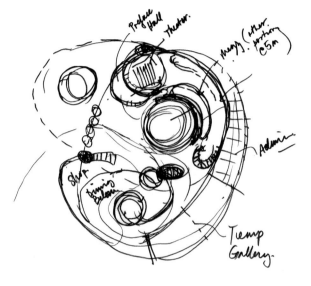

Astronomical Instruments

Three "celestial bodies" comprise major features within the architecture: the Oculus, the Inverted Dome and the Sphere. Each major element acts as an astronomical instrument, tracking the Sun, the Moon and stars and reminding us that our conception of time originates in distant astronomical objects. The building form, program and circulation further incorporate the orbital movement, supporting the flow of visitors through the galleries and the experience of the three central bodies.

天文功能

整座天文馆的三大核心节点——圆洞天窗、倒转穹顶和天象厅球体均有着天文学功能，共同诠释着日月星辰运行的基本规律，并提醒我们时间的概念起源于遥远的天文现象。馆内空间形态、展项设置和流线设计进一步与星体轨道运动相结合，参观者可以一边观看展览，一边体验三个主体核心节点。

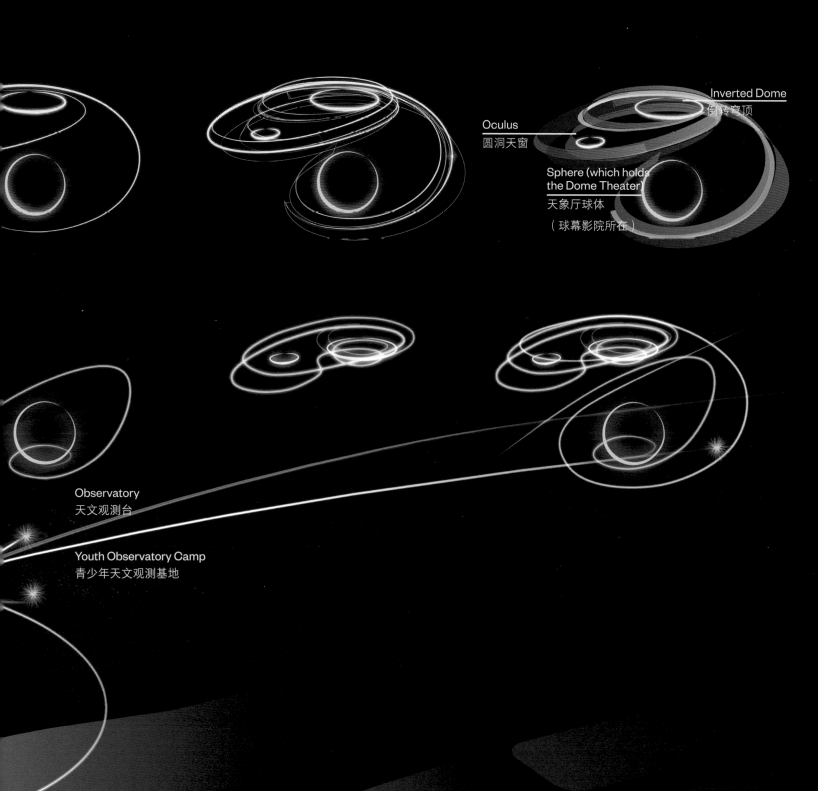

The Three Instruments
三大节点

Realization
实现

Modern life has detached most of us from that elemental bond. We take for granted the functional aspects of the Earth's daily rotation as it orbits around the Sun, our Moon's circumscribing watch around the globe—a reality that more and more often becomes filtered through an app on a tiny screen.

现代生活已经使绝大部分人脱离了这种纽带关系。昼夜的交替、月亮的盈亏、四季的更迭都被简化为手机屏幕或时钟程序上跳动的数字，再无法引起我们深究的兴趣。

— Thomas Wong
托马斯·黄

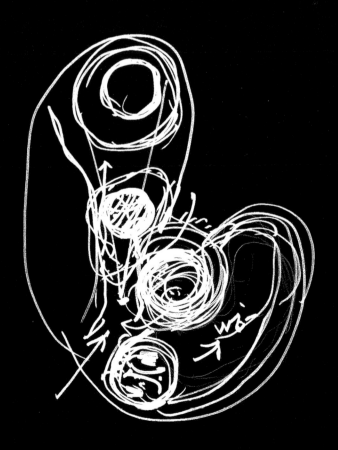

At 38,200 square meters, the Shanghai Astronomy Museum is the world's largest museum dedicated to astronomy. Forward-looking while presenting links to the past, it mirrors the rich history of Chinese astronomy and the future ambitions of the country's space exploration program.

The ancient Chinese concept of dualities, or Yin and Yang, is represented in multiple layers that are revealed to visitors as they move throughout the project: ancient versus future, earthbound versus floating, glass versus solid, building versus landscape, moving versus stationary.

Making reference to Chinese mythological forms— the Earth as square and the heavens as round— the building is dynamically positioned on the site: It is rooted asymmetrically on one side, while the rest of the form appears to vault toward the sky. A bold, monumental figure within the park, it embodies the aspirations of future astronomers and exploration into deep space.

上海天文馆占地5.86万平方米，总建筑面积3.82万平方米，是世界上最大的天文馆。在呈现与过去联系的同时，它还展示了中国天文学的悠久历史和中国未来探索太空的雄心。

天文馆将中国古代的阴阳二元概念在多个层面进行了呈现。参观者在整个场馆中走动时，会时时感受到这样的对比：远古与未来、落地与浮空、易碎与坚固、建筑与景观、运动与静止……

参照中国神话中"天圆地方"的理念，天文馆建筑被赋予了动感的不对称外形：其一侧稳坐于地面，其余部分则向上延伸。这个大胆、雄伟的形象，表现了未来天文学家的抱负和对深空探索的渴望。

The museum is dedicated to the popularization of the astronomical sciences, integrating exhibits and permanent collections with areas for observation and research. Included are permanent and temporary exhibit areas, an optical Dome Theater, a digital Dome Theater and auxiliary and educational programs. Three levels are above grade, and one is below.

上海天文馆致力于天文科学的普及，将展览与永久藏品区、观测与研究区相结合。馆内将设有永久和临时展览区、光学天象厅、数字天象厅，以及附属设施区和教育区。整个建筑包括地上三层，地下一层。

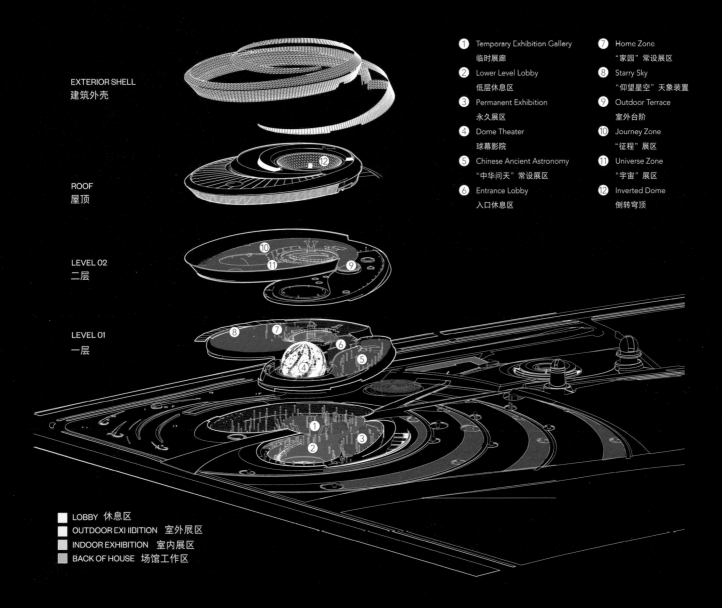

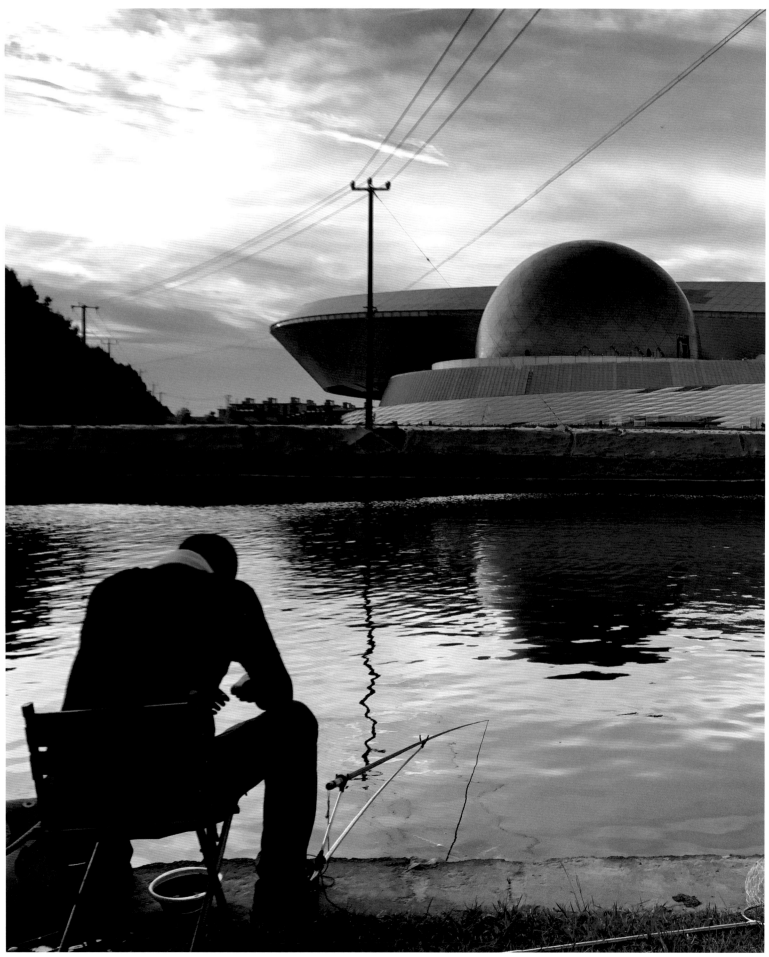

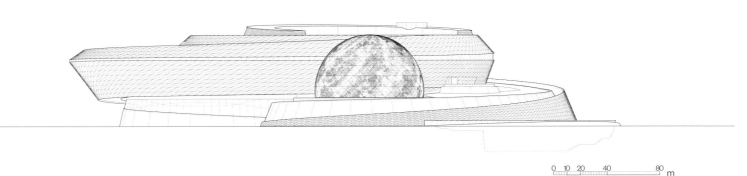

South Elevation 南立面

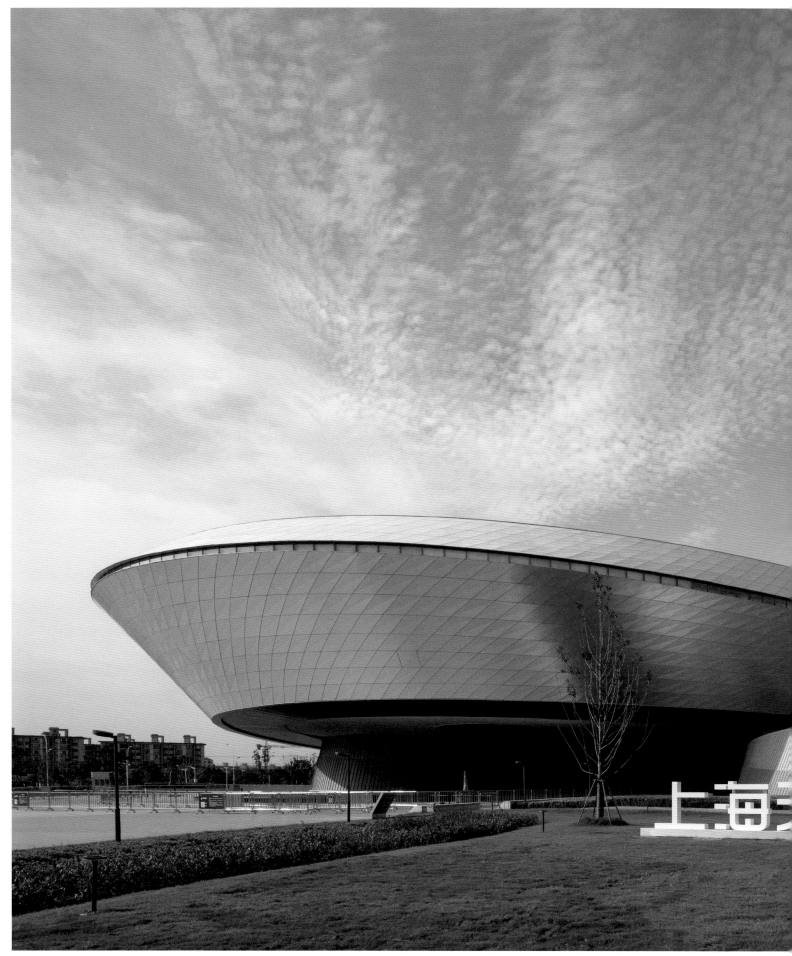

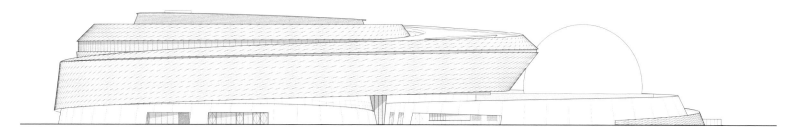

West Elevation　西立面

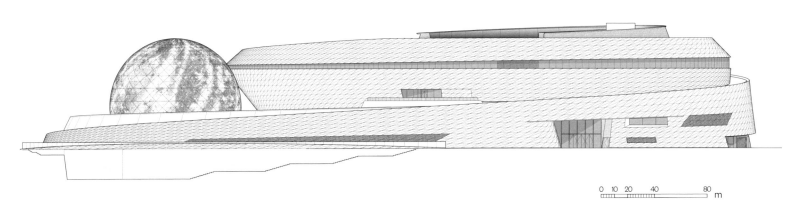

East Elevation 东立面

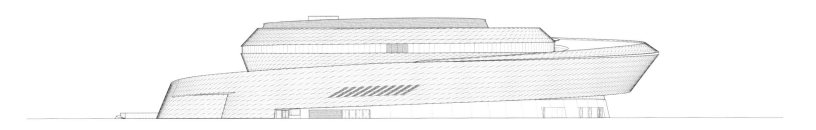
North Elevation 北立面

The Oculus
圆洞天窗

Suspended above the main entry to the museum, the Oculus demonstrates the passage of time by tracking a circle of sunlight on the ground across the entry plaza and reflecting pool. Shaped to synchronize with the change of the Sun's altitude over the course of the year, registration arcs at the plaza below the Oculus index the figure of light with both the time of day and time of year. In effect, the Oculus becomes a sun dial at the building scale. Like the prototypical clock tower in the town square, the Oculus is the timepiece of Lingang's new civic center.

圆洞天窗位于天文馆主入口上。阳光穿过圆洞时会在地面形成光斑，且随着太阳在天空中的移动，光斑也会随之在入口广场和反射池之间移动，以反映时间的流逝。夏至正午时分，光斑会与入口广场地面上的圆形标志完美重合，成为节气标志。整个圆洞天窗好似一个日晷，无时无刻不在捕捉光影，记录时间。就城市尺度而言，圆洞天窗如同市民广场上的钟楼，同时也成就了新的市民中心。

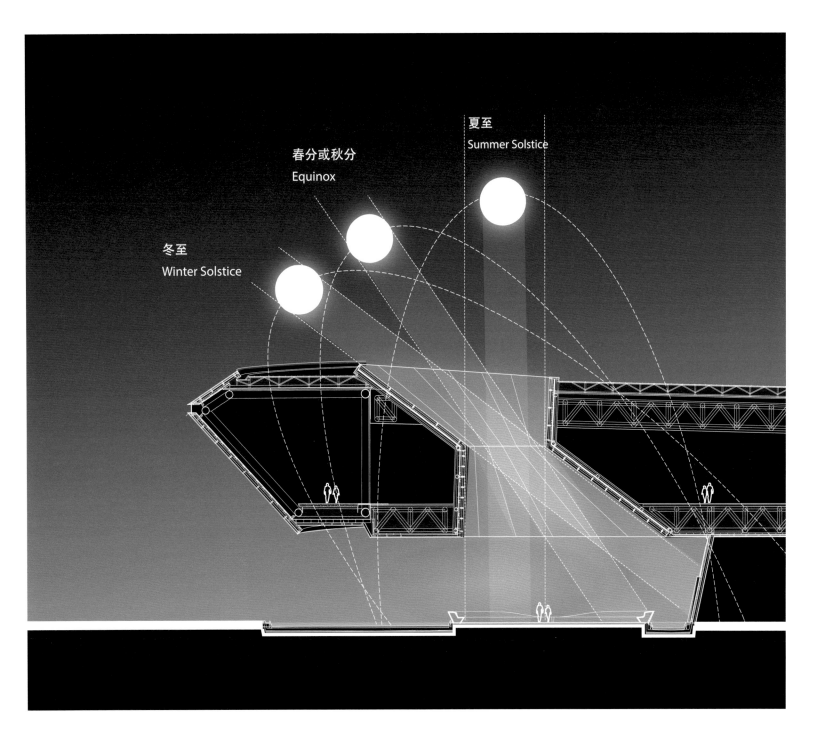

    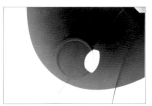

Solar timelapse study of the Oculus　圆洞天窗日照延时投影研究

Sunlight timelapse under the Oculus
从圆洞天窗落下的阳光的延时摄影

10:00 AM

12:00 PM

2:00 PM

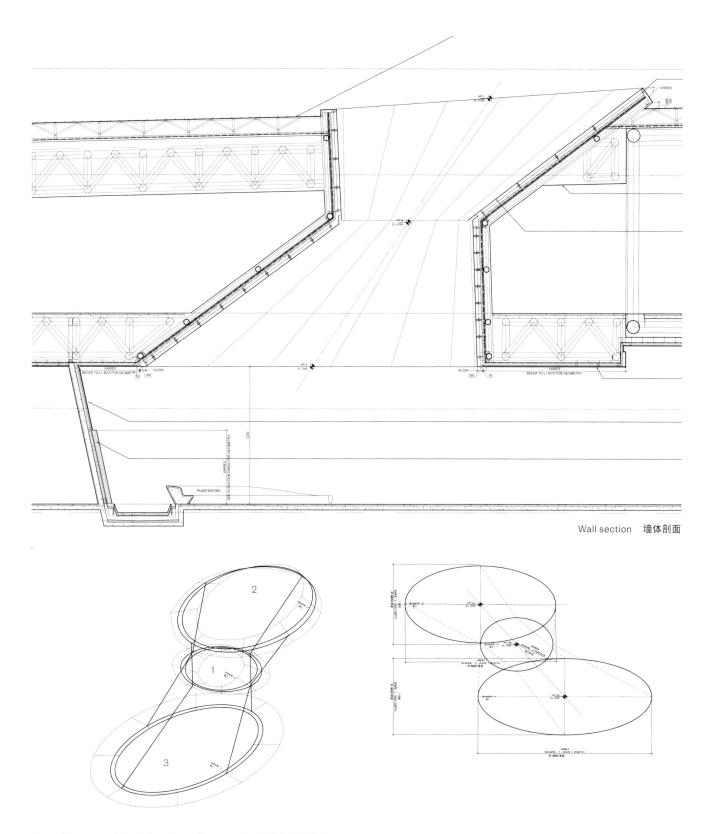

Wall section 墙体剖面

Shape 1 has a parallel relationship to the ground　图形1平行于地面
Shapes 2 and 3 have a perpendicular relationship to the Oculus　图形2和图形3垂直于圆洞天窗

The Atrium
中庭

The centerpiece of the museum and the heart of both the architectural and exhibit sequence is the main atrium, with its spiraling ramp and the Inverted Dome cradled by a massive concrete tripod. One emerges into this dynamic space immediately after the sense of compression through the entry procession under the building's cantilever. This set piece is the ever-present reference point for the visitor, the space which starts the experience, where one passes through in between galleries and contains the culmination of the exhibit sequence with a real exposure to the sky. Like the experience in some of the great cathedrals of the West, this space is a constant reminder of the dynamic power of the universe, an embodiment of orbital motion which propels the planets, and the Inverted Dome as the conclusion that directs one upward to the sky as the final act, as if to say: "Explore this!"

主中庭是天文馆的中心,也是建筑和展览序列的核心,其螺旋坡道和倒转穹顶由一个巨大的混凝土三脚结构柱支撑。从入口处的悬挑下方进入中庭,便能立即体验到空间由狭窄到豁然开朗的巨大反差。这里既是空间体验的开始,也是一个永恒的位置锚点,由此人们可以穿梭于各展廊之间,于苍穹之下饱览展览的精彩之处。类似于一些西方大教堂带给人的空间体验,这里不断向人们提示着宇宙的动态能量,并将行星的轨道运动予以具现,而倒转穹顶作为这场旅程的最后一站,将人们的视线引向苍穹,仿佛在说:"探索它!"

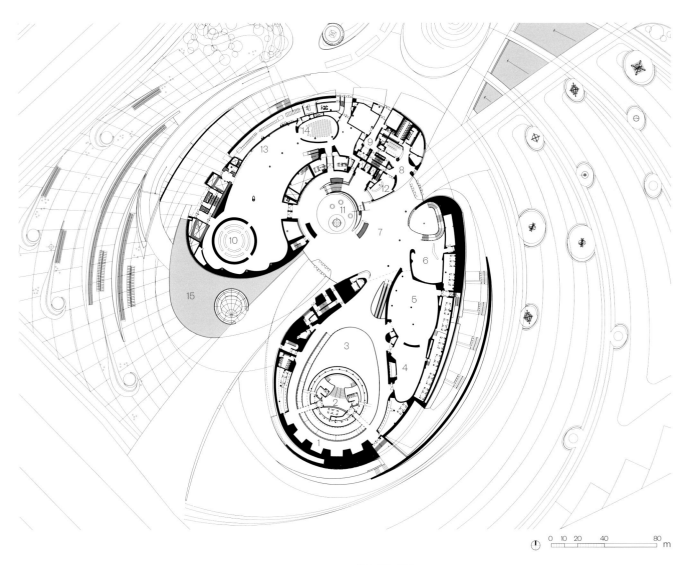

Ground Floor Plan

1. Gallery
2. Dome Theater
3. Dome Theater Lobby
4. Sail to Mars
5. Chinese Ancient Astronomy
6. Science Mall
7. Entrance Lobby
8. Coat Check
9. Museum Services
10. Starry Sky
11. Pendulum
12. News Interview Room
13. Home Zone
14. Flyover Galaxy Theater
15. Reflecting Pool

首层平面图

1 展廊
2 球幕影院
3 球幕影院休息区
4 "航向火星"常设展区
5 "中华问天"常设展区
6 科技大厅
7 入口休息区
8 衣物寄存处
9 天文馆服务区
10 "仰望星空"天象装置
11 钟摆装置
12 媒体采访区
13 "家园"常设展区
14 "飞越银河系"剧场
15 反射池

The 720-degree spiraling ramp inside the museum and underneath the Inverted Dome traces the orbital flow of the visitor sequence throughout the museum exhibits and launches the eye upward to its apex.

位于博物馆中庭、倒转穹顶下的720°螺旋坡道引导了整个观展流线，并将访客视线引向整个博物馆的最高点。

Visitors circulate throughout the structure in the manner of orbiting celestial bodies, bathed in constantly changing sculpted sunlight.

沐浴在不断变化的交错光影中,参观者们按照天体沿轨道运行的方式在馆内穿行。

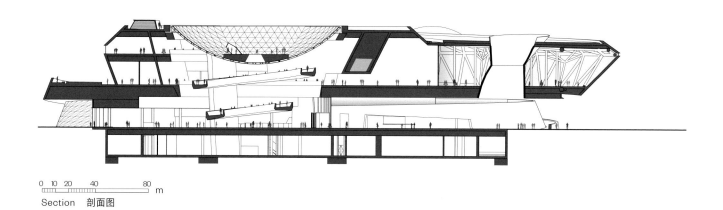

0 10 20 40 80 m
Section 剖面图

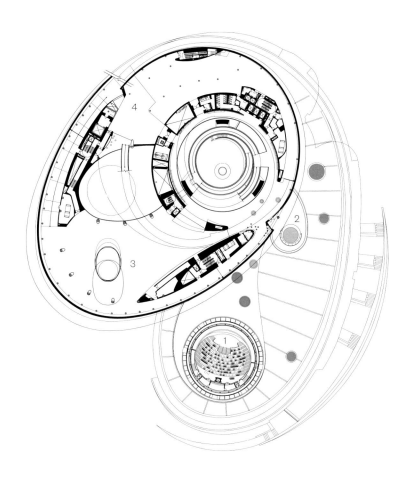

Second Floor Plan
1. Dome Theater
2. Outdoor Terrace
3. Universe Zone
4. Journey Zone

二层平面图
1 球幕影院
2 室外台阶
3 "宇宙"常设展区
4 "征程"常设展区

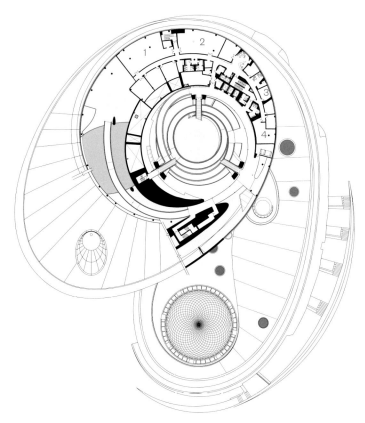

Third Floor Plan
1. Meeting Rooms
2. Offices
3. Executive Offices
4. Archive

三层平面图
1 会议室
2 办公区
3 管理办公区
4 档案室

The Sphere (which holds the Dome Theater)
天象厅球体（球幕影院所在）

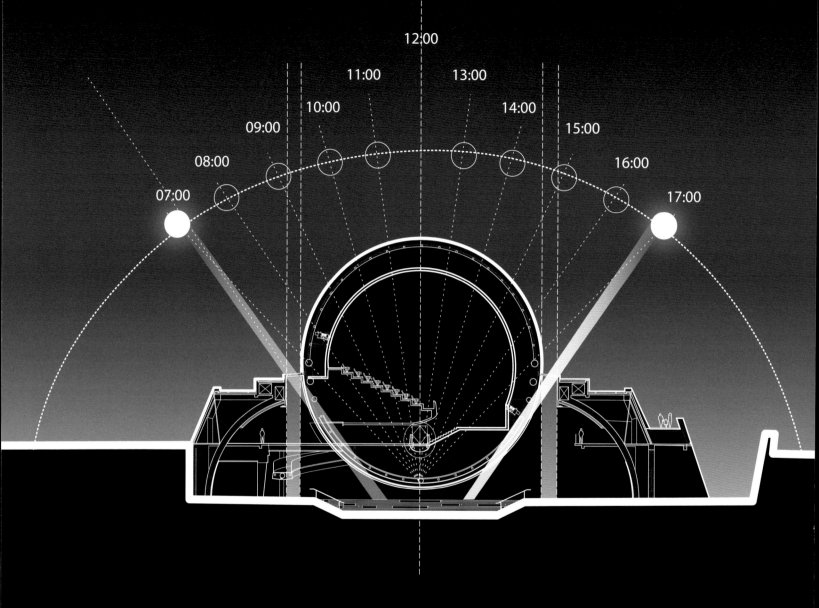

The Sphere contains the domed screen of the digital planetarium. A continuous skylight around the Sphere allows direct sunlight to enter and marks the passage of time in the museum below, with a full circular ring of light realized at the noon hour of the summer solstice. The Sphere derives its shape not only from this programmatic element but also as a manifestation of the primary celestial forms. The Sphere's apparent suspension in its roof structure allows visitors to experience it as a weightless mass from below. The Sphere is perceived as an ever-present reference point, like the Sun, in the visitor experience.

球幕影院位于天象厅球体的球状空间内。一系列环形天窗紧绕中心球体，阳光透过天窗射入天文馆内部，形成了展示时间流逝的印记，并在夏至午时与地面圆环完全重合。球体设计不仅代表了功能分区的一部分，同时也是对天体形式的具象化展现。悬挂式的球体呈现策略使得游客在下方行走时拥有了不一样的失重体验，这是一个如同太阳般的、整个观展流线的焦点和中心。

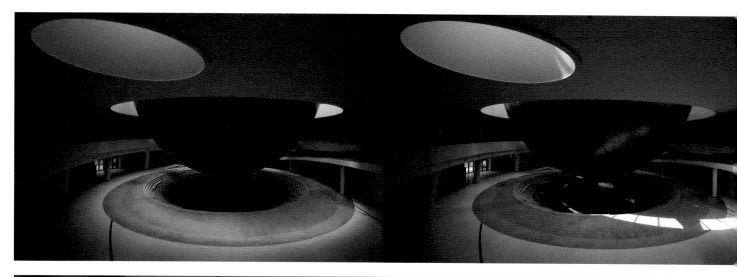

Time-lapse of the Sphere (which holds the Dome Theater) at the summer solstice
夏至日天象厅球体（球幕影院所在）的延时摄影照片

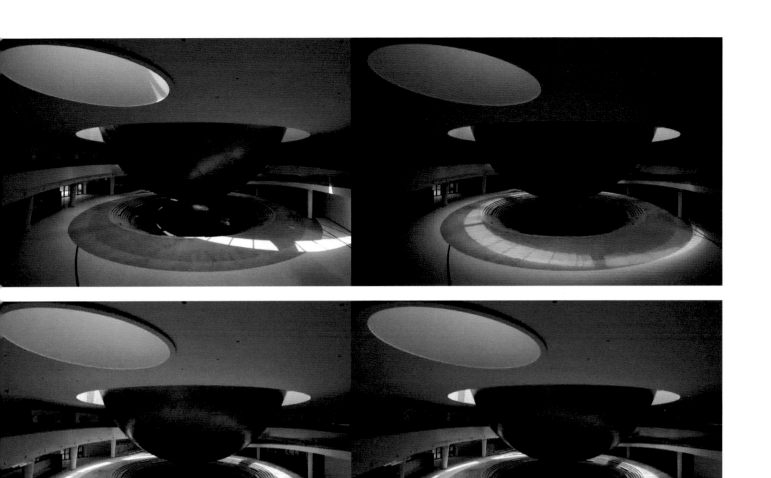
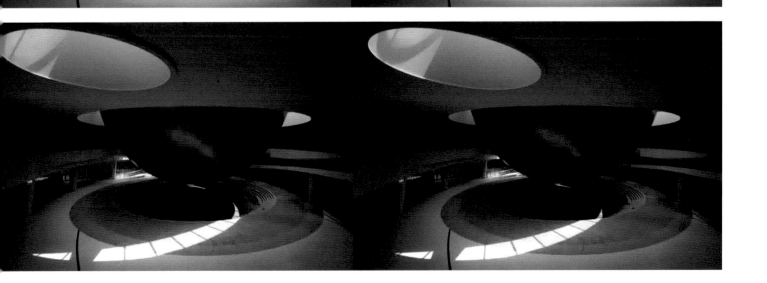

A pattern of indentations on the surface of the Sphere represents the topography of an abstract celestial body, as seen from a distance in space.

球体表面的材料纹理模拟了从太空中观察到的某一颗抽象天体的地貌。

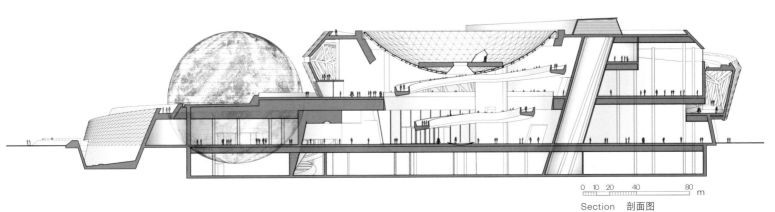
Section 剖面图

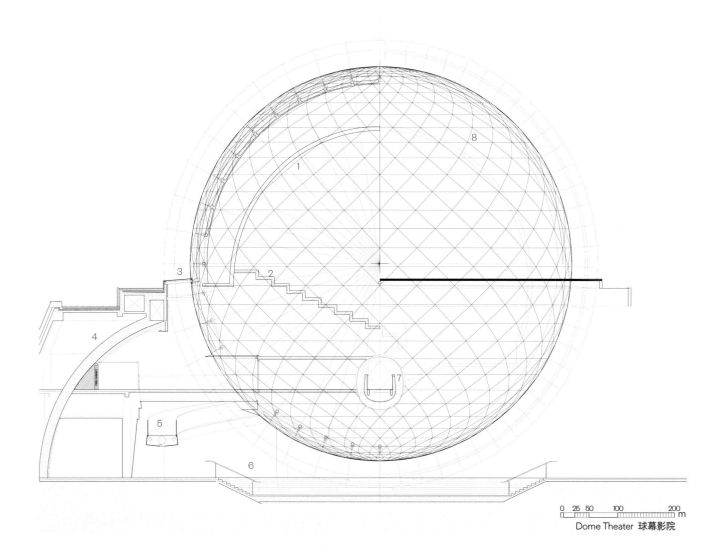

Dome Theater 球幕影院

The Sphere Section
1. Interior Domed Projection Screen
2. Dome Theater Seating Risers
3. Continuous Glass Skylight
4. Structural Concrete Dome
5. Ramp from Main Level to B1 Level
6. Seating area under sphere
7. Entry bridge to Dome Theater
8. Sphere metal panel cladding pattern

天象厅球体剖面
1 室内穹顶投影屏
2 球幕影院阶梯座位区
3 连续的玻璃天窗
4 结构混凝土穹顶
5 主层至B1层坡道
6 球幕下方的座位区
7 球幕影院入口连桥
8 带纹理的球形金属挂板

The 23-meter-diameter Dome Theater is housed within the Sphere, which is half submerged in the building. The theater utilizes an advanced, high-definition spherical projection system and laser performance system.

直径23米的球幕影院位于球体内部，采用了先进的高清球面投影系统和激光表演系统。

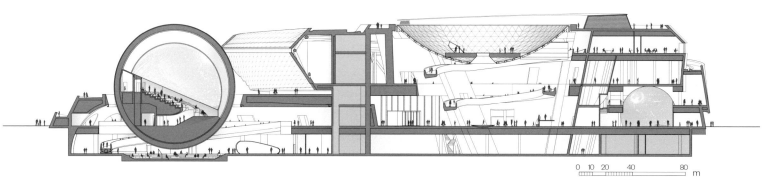

Section 剖面图

Exhibit Galleries
展览馆

Through the museum's exhibits and experiences, visitors gain a holistic understanding of the universe and the connection between humankind and the cosmos. Adults and children can delve into the Big Bang, the formation of the first particles, the creation of our solar system and planets—leading them to a greater understanding of and respect for the origins of life on Earth. The exhibits also provide an up-to-date look at contemporary Chinese and Western space exploration and its cultural implications—and pique visitors' curiosity by contemplating the unsolved mysteries ahead as humans continue to make strides in the astronomical sciences.

上海天文馆精心的展项设置与体验策划，使参观者能够对宇宙以及人类与宇宙的联系建立全面的了解。各个年龄层的观众都可以在这里深入了解宇宙大爆炸、第一批粒子的形成、太阳系及其行星的形成等天文知识，从而对地球上生命的起源产生更深刻的理解与敬畏。同时，天文馆内还呈现了一些当代中国和西方空间探索的最新成果及其文化含义，吸引参观者共同思考那些随着天文学发展而不断涌现的未解之谜。

Ambient atmospheres, simulations and lighting and sound effects create an immersive environment in the exhibition areas. Advanced displays enable sensory interaction, data visualization, augmented and virtual reality and biometric experiences.

在展览区域，环境气氛、拟真和声光效果创造出沉浸式的体验环境。先进的显示设备使感官交互、数据可视化、增强现实、虚拟现实以及生物识别都成为现实。

The exhibition houses more than 70 meteorite samples, including specimens that originated from Mars, the asteroid Vesta, and the moon. A range of cultural relics include original work by Galileo Galilei, Tycho Brahe, Johannes Kepler, Isaac Newton, and other astronomers.

天文馆共展出了来自包括火星、灶神星及月球的70余个陨石标本，以及人类历史上重要天文学家伽利略、第谷、开普勒、牛顿等人的工作记录，它们是人类的文化遗产。

The Inverted Dome
倒转穹顶

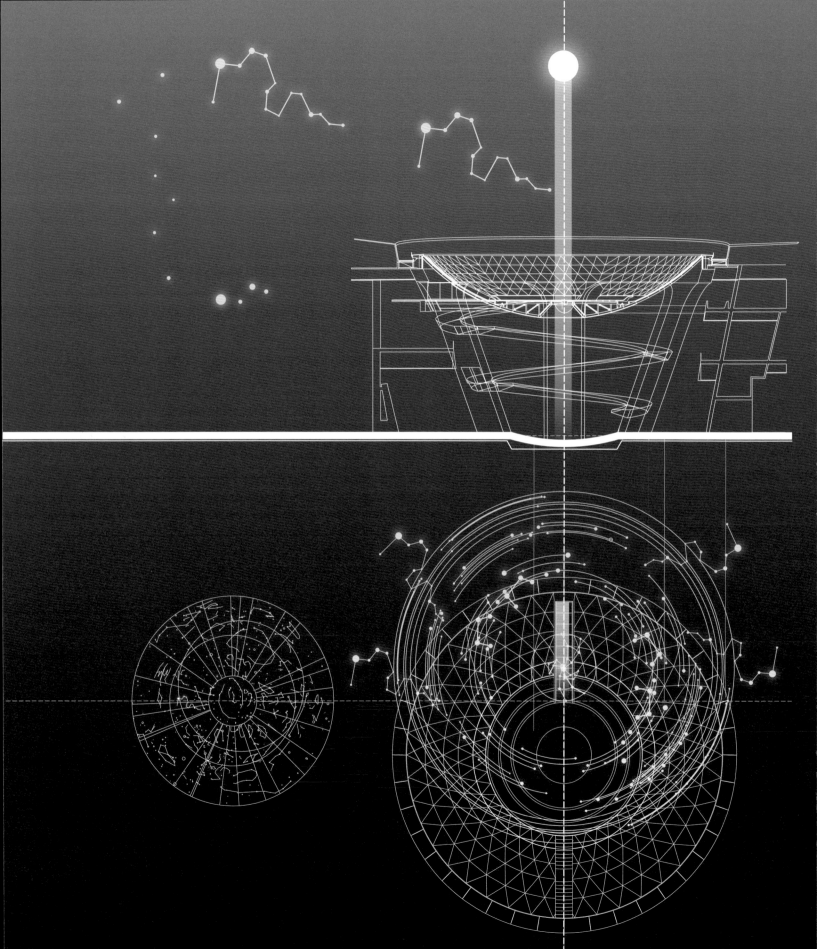

The Inverted Dome is the culmination to the visitors' exhibits experience. A large inverted glass tension structure, it sits atop the central atrium at the roof line. Visitors can occupy the center of the glass dish for an unimpeded view of the sky—a true encounter with the universe to conclude the simulated experience within the museum.

倒转穹顶是整个观展体验的高潮。巨大的、极富戏剧张力的玻璃穹顶结构位于整个建筑中庭屋顶中央。当访客置身于穹顶中心时，一览无遗的天空映入眼帘：这是访客与宇宙真正的、最直接的相遇，也是整个参观的高潮与华彩。

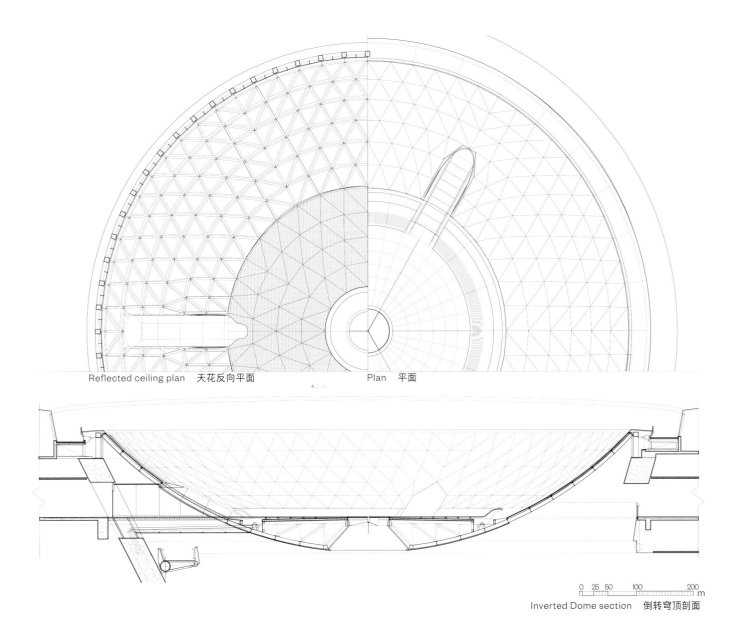

Reflected ceiling plan 天花反向平面   Plan 平面

Inverted Dome section 倒转穹顶剖面

The Inverted Dome, which facilitates an authentic experience of both day and night sky, sits atop the central atrium around which all galleries are organized and through which all visitors pass. Restricting the view of the neighboring context by eliminating the horizon, the Dome focuses visitors toward the heavens and becomes the real counterpart to the virtual star show visitors will experience in the planetarium.

倒转穹顶位于中央中庭顶部，无论昼夜均能提供极佳的天空视野，所有的展厅都围绕此中庭搭建。穹顶将周围的地平线遮挡起来，使参观者的注意力聚焦在天空，他们在天文馆虚拟星空展览内所看到的景象将真实地在这里展现。

Engineering and Construction
工程与建造

The realization of architecture is a seamless process from initial concept to opening day, and a successful design cannot be separated from its execution. To create a building such as this one—tuned to astronomical phenomena, that functions as a timepiece tracking the course of the Sun in the sky, integrating heroic forms—requires exceptional dedication, focus and precision throughout the construction phase. As architects, it is a truly moving experience to see ideas which first began to take shape in the mind materialize in physical form during construction. Bringing the Shanghai Astronomy Museum to fruition involved the efforts of so many across the span of several years, and its birth was a spectacular event to witness.

一个建筑项目的实现，从最初概念到建成开放的每个环节均不可或缺，成功的设计离不开它的执行。要打造出这样一座具有史诗般的外观、能够对天文现象进行动态回应的建筑，需要在整个建造阶段都保持非凡的奉献精神、专注力和对精确性的追求。作为建筑师，看到最初脑海中的想法随着施工的推进而逐渐成为现实，心中的感动难以言表。上海天文馆的建成，历经数年与多方的努力，其诞生是一件值得见证的盛事。

— Thomas Wong
托马斯·黄

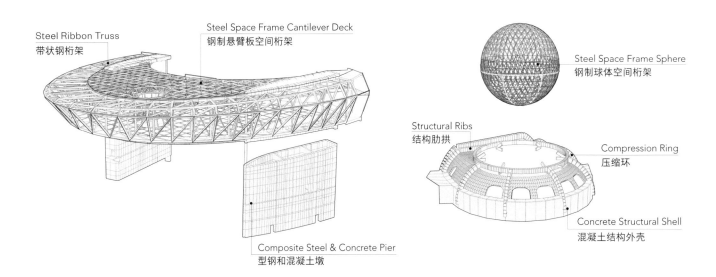

Steel Ribbon Truss
带状钢桁架

Steel Space Frame Cantilever Deck
钢制悬臂板空间桁架

Steel Space Frame Sphere
钢制球体空间桁架

Structural Ribs
结构肋拱

Compression Ring
压缩环

Composite Steel & Concrete Pier
型钢和混凝土墩

Concrete Structural Shell
混凝土结构外壳

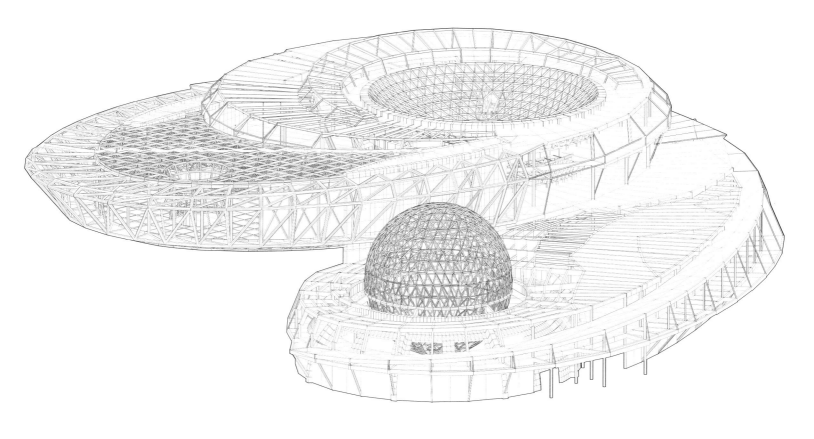

The main building is a hybrid structure of concrete and steel. The steel-framed Sphere, which holds the Dome Theater, is suspended within a concrete dome that spans 61 meters. A hole and ring beam at the top connects to the Sphere with six steel brackets, giving the illusion of a levitating body.

The 36-meter cantilever at the nose of the museum was achieved through an arching belt truss that also forms the outermost spiral of the building massing. The belt truss is connected to two massive concrete piers that serve as both building infrastructure (stairwells and shafts) and superstructure. The main floor and roof of the second-floor gallery are a three-dimensional space frame designed to minimize interior columns.

上海天文馆主体建筑为钢筋混凝土结构。球幕影院所在的天象厅球体悬挂于跨度为61米的大悬挑穹顶内。建筑顶部的穿孔与环梁通过6个钢结构支架将穹顶与球体相连，营造了一种球体悬浮的错觉。

天文馆前端长达36米的大悬挑形体则通过拱形带状桁架实现。该桁架也构成了建筑最外层的螺旋结构。带状桁架将两个巨型混凝土墩相连，形成了建筑的上层结构及容纳基础设施的功能区间（即楼梯间和管井）。二层展厅主体空间与屋顶部分形成了一个三维立体的框架结构，并最大限度减少了空间内部梁柱的使用。

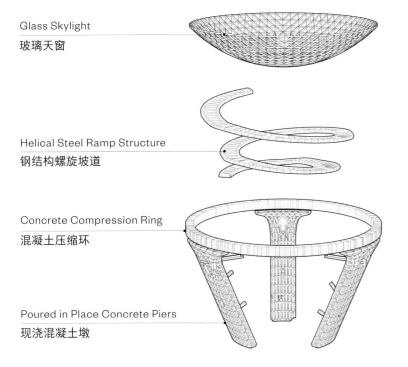

Glass Skylight
玻璃天窗

Helical Steel Ramp Structure
钢结构螺旋坡道

Concrete Compression Ring
混凝土压缩环

Poured in Place Concrete Piers
现浇混凝土墩

The Inverted Dome is a glass-clad tension structure that holds an occupiable platform at its base. The dome is held up by an asymmetrical tripod of steel and concrete piers that are tied at the top with a continuous compression ring. The piers also support the 720-degree spiraling ramp that descends from the Inverted Dome to the main atrium floor.

倒转穹顶是一个玻璃张拉结构，其底部有一个可进入的平台。穹顶由一个钢和混凝土支柱组成的不对称三脚架结构支撑，顶部用一个连续的压缩环连接。该三脚架结构还支撑着连接倒转穹顶与主中庭地面的720°螺旋坡道。

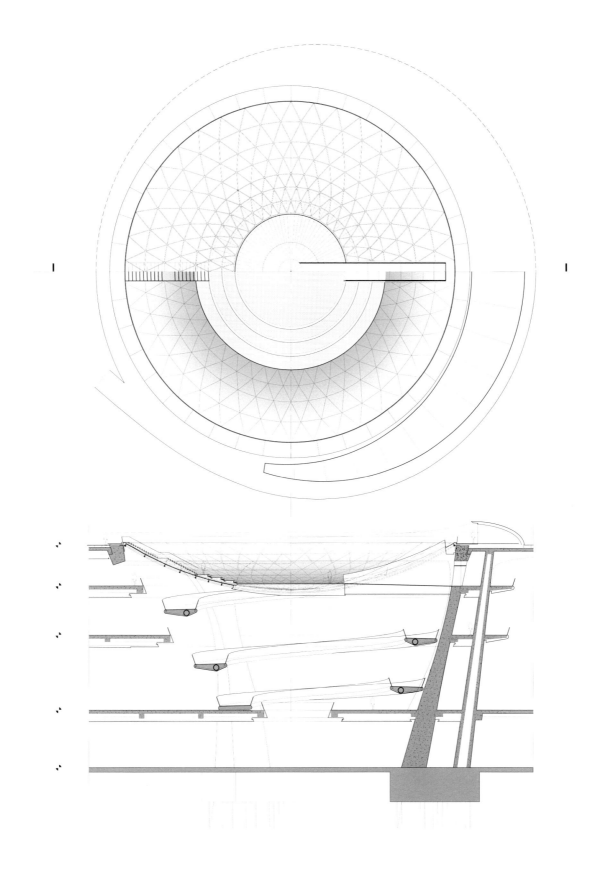

December 2017 二〇一七年十二月   December 2017 二〇一七年十二

September 2018 二〇一八年九月   September 2018 二〇一八年九

December 2018 二〇一八年十二月   March 2019 二〇一九年三

September 2018　二〇一八年九月

September 2018　二〇一八年九月

September 2018　二〇一八年九月

November 2018　二〇一八年十一月

June 2019　二〇一九年六月

July 2019　二〇一九年七月

A concrete dome structure supports the steel-framed Sphere. A series of apses are cut out of the concrete dome at the main level to create exhibition spaces and serve as the building's "ambulatory," or processional as seen behind the altar in Gothic cathedrals.

混凝土穹顶结构支撑着钢桁架球体。一系列壁龛从主层的混凝土圆顶中剪切割出，以创造展览空间，并作为建筑物的"回廊"，类似于哥特式教堂里祭坛后面的递进式空间。

Construction Site of the Inverted Dome 倒转穹顶施工现场

"I distinctly remember one visit while the atrium was still under construction, where the sheer muscle of the concrete tripod, the beautiful streaming light emanating from above, the steel skeleton of the spiraling ramp were all clearly evident. These features were then combined with the absolutely raw state of construction progress, the atrium filled with partially dismantled steel scaffolding and sparks flying through the air from a welder. I was left speechless at that moment, as well as much of the rest of the day. It was then that I understood the potential power of this building."

"我清楚地记得有一次前往现场,当时中庭仍在建设中。混凝土三脚架的纯粹结构、从上方倾泻而下的耀眼天光,以及螺旋坡道的钢骨架都清晰可见;与此形成鲜明对比的是施工现场那原始而粗犷的氛围——中庭堆满了尚未完全拆除拆卸的钢制脚手架,电焊火花在空中飞舞。那一刻情景给我的震撼无以言表,直到那天深夜仍令我心中激荡。就在那时,我明白了这座建筑的潜在力量。"

— Thomas Wong
托马斯·黄

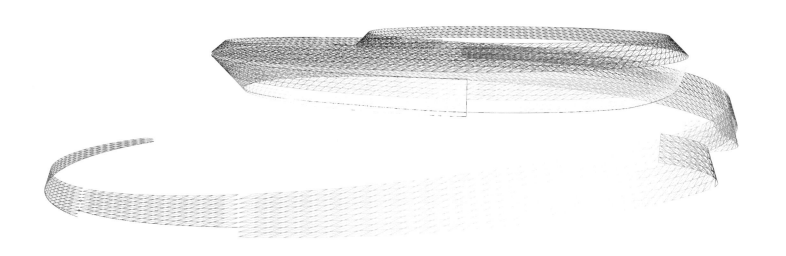

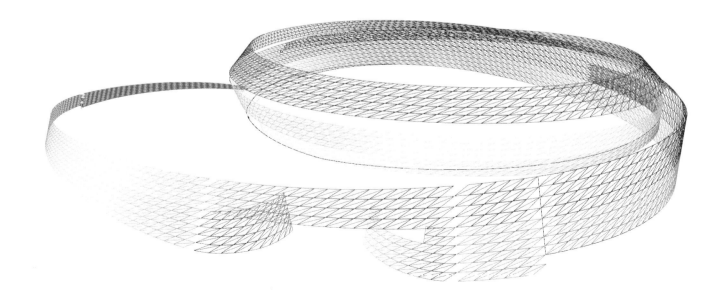

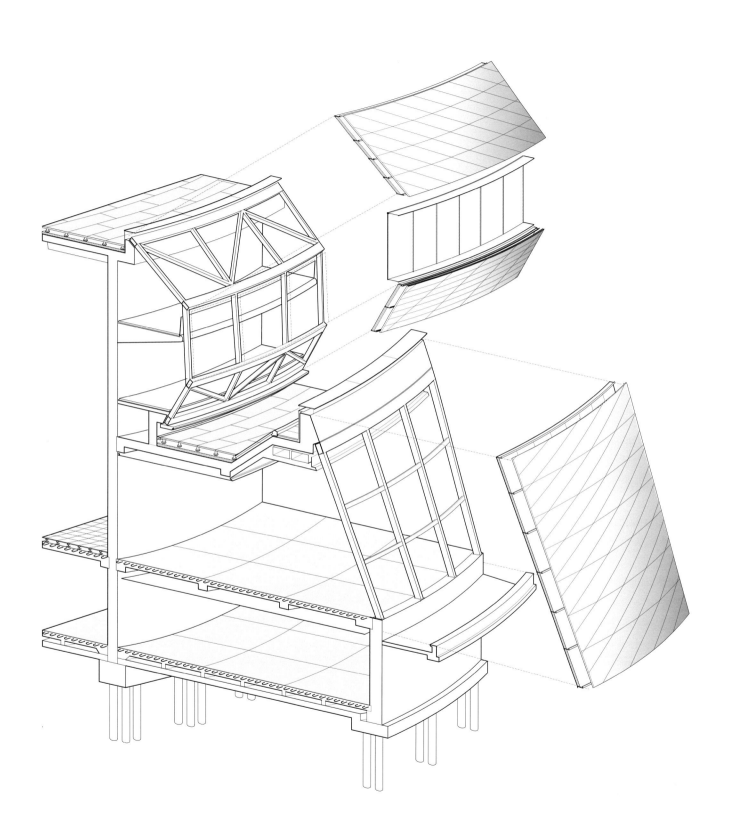

Diagram of skin system　表皮系统分析图

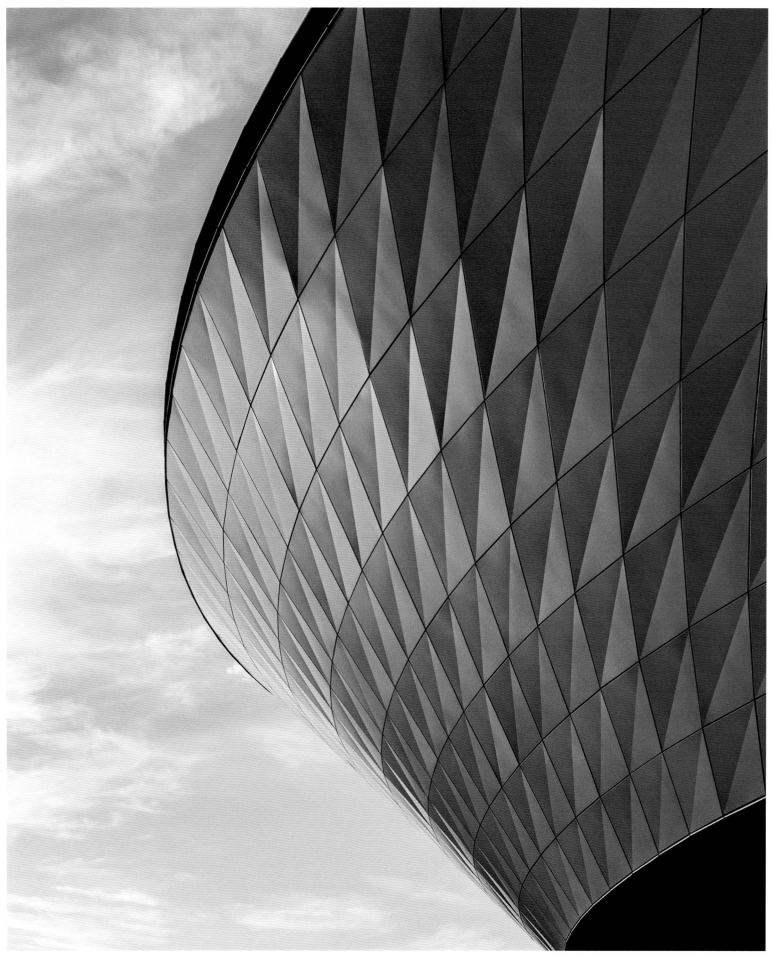

Acknowledgements

Ours is a profession of the collective whose achievements cannot possibly be reached without the vision, talents and skills of countless contributors. There are too many to cite all individually; but the special efforts of some cannot go unmentioned. Our sincere thanks:

To Dr. Ye Shuhua from Shanghai Astronomical Observatory, Chinese Academy of Sciences; Wang Lianhua, Wang Xiaoming, Xin Ge, Gu Qingsheng, Jin Jianmin, Xu Xiaohong, Lin Qing and Wu Qun from Shanghai Science and Technology Museum.

To our many professional partners without whom we could not have successfully accomplished the design of this building: Liu Enfang, Chen Yizhi, Liang Fei, He Tao, Zhang Lianglan from SIADR; Dan Sesil, Winnie Kwan from LERA; Wai Mun Chui from Brandston Partnership; Alexander Brandt and the Xenario Exhibit Design team; Jim Sweitzer; Ralph Appelbaum and the team at RAA for all the efforts during the competition phase; and our photographers: Arch-Exist Team, Yihuai Hu and etc.

To the myriad contractors, subcontractors, suppliers and construction team, whose day-by-day coordinated teamwork brought this vision to reality for the world to experience.

To the talented members of the Ennead team, whose dedication and brilliance are embodied in the excellence of our work.

Ennead Architects LLP

致谢

建筑行业是一个集体性的行业，若没有无数贡献者的远见、才华和技能，这些成就便无以实现。虽无法向所有人员一一致谢，但在这里，我们希望至少向下列人员的特别付出致以衷心的感谢：

上海天文台叶叔华院士；上海科技馆王莲华、王小明、忻歌、顾庆生、金建敏、徐晓红、林清、吴群。

来自上海建筑设计研究院有限公司的刘恩芳、陈逸芝、梁飞、何涛、张良兰；来自LERA的Dan Sesil、Winnie Kwan；来自Brandston Partnership的Wai Mun Chui；Alexander Brandt和飞来飞去展览设计团队；Jim Sweitzer; Ralph Appelbaum和RAA团队在国际竞赛阶段提供的所有帮助；以及我们的摄影师们：存在建筑团队、胡艺怀等。没有诸位专业合作伙伴，我们就无法圆满完成这座建筑的设计。

感谢所有承包商、分包商、供应商和施工团队，是你们日复一日的协调合作使这座天文馆从愿景变为现实，呈现在世人面前。

最后，感谢艺艾德团队的各位成员，这座建筑的成功离不开你们付出的辛勤与才智。

艺艾德建筑设计事务所

Museum Client 委托方	Shanghai Science and Technology Museum 上海科技馆
Design Architect 建筑设计	Ennead Architects LLP 艺艾德建筑设计事务所 Design Team 设计团队 Design Partner 设计合伙人：Thomas Wong Management Partner / Principal 管理合伙人 / 项目负责人：V. Guy Maxwell / Grace Chen Project Manager 项目经理： Weiwei Kuang Project Designer 项目设计师：Charles Wolf Project Architect 项目建筑师：Anthony Guaraldo Design Team 设计团队：Jorge Arias, Margarita Calero, Michael Caton, Christina Ciardullo, Eugene Colberg, Regina Jiang, Jörg Kiesow, Aidan Kim, Stefan Knust, Xinya Li, Francelle Lim, Xiaoyun Mao, David Monnar, Nikita Payusov, James Rhee, Yong Roh, Miya Ruan, Na Sun, Eric Tsui, Stephanie Tung, Charles Wong, David Yu, Fred Zhang
LDI (Local Design Institute) 当地合作设计院	Shanghai Institute of Architectural Design & Research (Co., Ltd.) Architectural Design Department 5 上海建筑设计研究院有限公司 建筑设计五院
Exhibit Design 展陈设计	Xenario 上海飞来飞去展览设计工程有限公司
Lighting Consultants 照明顾问	Brandston Partnership Inc. (BPI) 上海碧甫照明工程设计有限公司
Code / Life Safety / Landscape / MEP / Structural 规范/消防/景观/机电/结构：	Shanghai Institute of Architectural Design & Research (Co., Ltd.) 上海建筑设计研究院有限公司
General Contractor 总承包商	Shanghai Construction No. 7 (Group) Co., LTD 上海第七建设（集团）有限公司
Book Designers 书籍设计	Thomas Wong, Aislinn Weidele
Editor 编辑	Carolyn Horwitz
Contributors 撰稿	Grace Chen, Keristen Edwards, Regina Jiang, Keri Murawski, Charles Wolf, Luyi Yu
Photography 摄影	Arch-Exist , Yihuai Hu, Justin Woo, Yaping Qi (Seven Panda)

图片出处
取材自NASA的太空摄影
P28 天鹅座面纱星云
P30 仙后座的泡泡星云
P32 哈勃太空望远镜拍摄的木星
P34 M87星系中心黑洞
P36 从国际空间站看到日出时的金星

Images Sources
Space photography courtesy of NASA
The Veil Nebula, in the constellation Cygnus
The Bubble Nebula, in the constellation Cassiopeia
Jupiter, photographed by NASA's Hubble Space Telescope
Black hole in Galaxy Messier 87
Venus at sunrise, as seen from the International Space Station

其余图片来自
新华社/Alamy图片摄影机构
视觉中国/财新网
詹姆斯·特瑞尔的作品 ©Siebe Swart
天坛 ©Access China
北京天文馆 ©N509FZ

Additional Images
Xinhua / Alamy Stock Photo
VCG / Caixin Global
James Turrell's works ©Siebe Swart
Temple of Heaven ©Access China
Beijing Planetarium ©N509FZ

About Ennead Architects

Ennead Architects is an internationally acclaimed architecture firm with offices in New York City and Shanghai that is renowned for its innovative design work across every building typology. Focused on form and function, Ennead demonstrates technical and artistic excellence that promotes civic engagement and cultural connection within communities.

Ennead works on a diverse folio of projects with clients both public and private: from Stanford to Yale universities, from the American Museum of Natural History to the William J. Clinton Presidential Center, from Carnegie Hall to the Standard Hotel, High Line. The practice has evolved continuously since 1963, resulting in hundreds of transformative spaces around the world. Regardless of project type, Ennead's team of 200+ design thinkers create expressive architecture that embodies each client's mission and engages the user, while shaping the public realm.

The firm's body of work—including new construction, renovation and expansion, historic preservation, interior design and master planning—ranges from museums and performing arts centers to complex laboratory, research and teaching facilities for higher education, to residential, commercial and hospitality projects, to large scale civic and infrastructure projects. Ennead is the recipient of numerous design awards, including the prestigious Smithsonian Institution-Cooper Hewitt National Design Award, the AIANY Medal of Honor, and the National AIA Firm Award. Nine partners lead the firm: V. Guy Maxwell, Kevin McClurkan, Molly McGowan, Richard Olcott, Tomas Rossant, Todd Schliemann, Peter Schubert, Don Weinreich, and Thomas Wong.

www.ennead.com

关于艺艾德建筑设计事务所

艺艾德建筑设计事务所总部位于美国纽约市，并在上海设有办公室，经过半个多世纪的发展，艺艾德已经发展成为建筑设计领域公认的、领先的国际化设计事务所。从形态，到布局，再到材质，艺艾德的建筑作品极具表现力和感染力，同时恰到好处地展现了客户的愿景，并且塑造了我们的空间和时代。

从斯坦福大学到耶鲁大学，从美国自然历史博物馆到克林顿总统纪念图书馆，从卡内基音乐厅到纽约高线公园标准酒店，艺艾德与诸多国际知名学术文化机构均有长期稳定的合作伙伴关系。自1963年成立以来，艺艾德的200余名建筑师即思想家们，始终致力于创造极具公众影响力的建筑作品，重塑公共建筑与社会以及人的关系，为来自公共机构和商业领域的众多客户创造了多个令人瞩目、富有创新性并秉承可持续发展理念的地标建筑作品。

艺艾德的设计服务内容包括新建建筑、扩建改建、历史保护、室内设计及总体规划，项目类型多元化，包括博物馆、表演艺术中心、高精尖实验室、高等教育研究设备、商业办公和大型基建项目等。艺艾德亦在建筑设计界屡获殊荣，包括著名的史密森协会-库珀·休伊特国家设计奖、美国建筑师协会纽约州分会主席奖、美国建筑师协会颁发的年度设计师事务所奖等。九名设计合伙人分别为：V. Guy Maxwell、Kevin McClurkan、Molly McGowan、Richard Olcott、Tomas Rossant、Todd Schliemann、Peter Schubert、Don Weinreich和Thomas Wong。

www.ennead.com

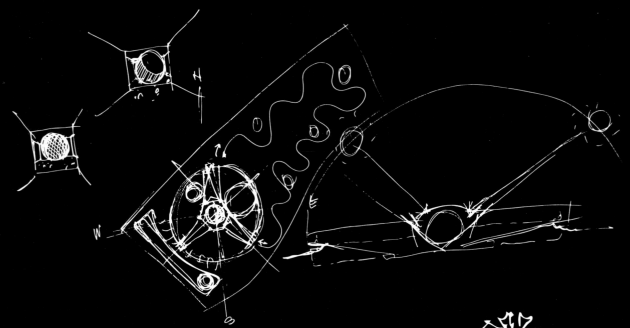

40 teenager
cabins
a modular craft
compact.

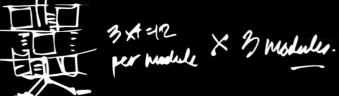

3x4=12
per module x 3 modules

40 = 8x5
40 = 2x10

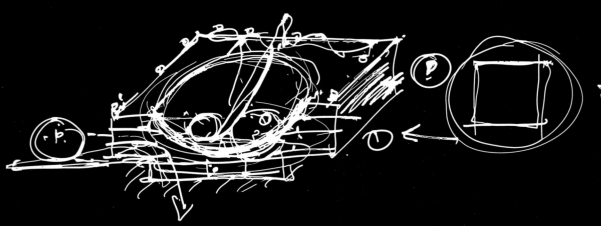

Temple of Heaven